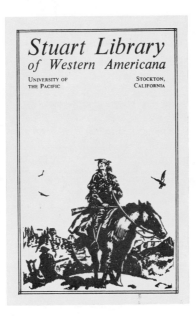

SPAIN IN THE WEST
XII

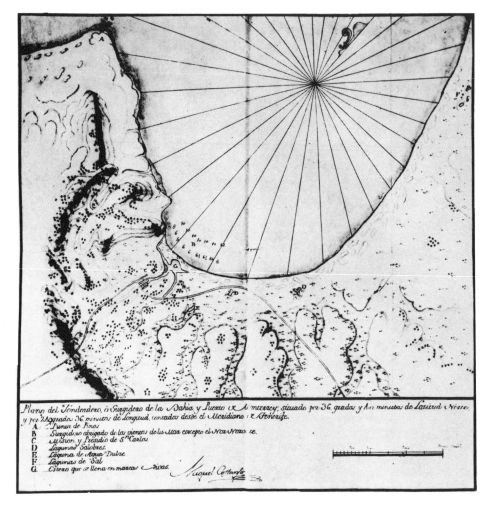

The Spanish
Royal Corps of Engineers
in the
Western Borderlands

Instrument of Bourbon Reform

1764 to 1815

by
JANET R. FIREMAN

THE ARTHUR H. CLARK COMPANY
Glendale, California
1977

to
BERT M. FIREMAN
the first historian I ever knew

Contents

Illustrations

Foreword

Years ago when I was first being introduced to the study of the Spanish Borderlands, I came across an occasional reference to a group of explorers or a party of settlers who were accompanied by the engineer Señor Fulano, or Lieutenant Zutano, or even Captain Menguano. That was it. No mention was found, not even embalmed in the most obscure footnote, as to who these engineers were or where they had come from; neither was there any inkling as to their specialization, whether civil, topographical, mechanical, sanitary or domestic. In further reading I gained few clues as to their origins or their reason for being present at such auspicious occasions as the history books noted. I could discover little of them either as persons or as part of any larger unit to which they might have belonged. It would have been easy to have conceded that they just appeared; but my curiosity was eventually rewarded, for the seemingly ever-present engineer Miguel Costansó had drawn maps of my native California, and they were good ones. But why and for whom had his efforts been expended?

Recent scholarship is adding significant dimensions

to Borderlands history and pushing back the frontiers of ignorance, particularly in the neglected area of the military contribution to the northern frontier of the Viceroyalty of New Spain. In the study which follows, Dr. Janet Fireman has brought out in documented detail the significant activity of the Royal Corps of Engineers in the western Borderlands, and has provided information on the origins and nature of the work accomplished by that group in the eighteenth century. An elite military corps comes to life in this study. Dr. Fireman's work, combined with other recent scholarship, provides us not only with the high points which earlier attracted historical attention, but also with the ups and downs of frontier military service. We are introduced to the disappointment of defeat and the satisfaction of achievement of real people acting in a real, often harsh environment.

The Royal Corps of Engineers was both a product of the Bourbon era of Spanish colonial control and the result of a longstanding, unmet need. This hitherto historically neglected task force "provided indispensable technical expertise to civilian and military officials for defense and extension of the frontier." Born along with the reform spirit which attempted a regeneration of a decaying colonial empire, the Corps can be seen in action in the Borderlands along with other groups which carried out the modified colonial policy of Spain, such as the Naval Department of San Blas, the Catalonian Volunteers, and the Intendency System, though the Corps predated all of these.

There was little romance attached to frontier military operations in the Provincias Internas, but there was strong positive impact on the future of that area in the inspections made by military engineers, and in the recommendations that they made for alteration and improvement of that expansive northern defensive line. The Corps of Engineers was the only elite unit, consisting of officers only, assigned to the Viceregency of New Spain. Its services were in far greater demand than its paucity of numbers could hope to supply. Furthermore, the Corps was not exclusively destined for frontier service, nor even principally so; to the contrary, the engineers responded to apparent need wherever it might occur. As individual engineers spent considerable time in the far north, as Dr. Fireman so clearly points out, it was because viceregal authorities considered their services there to be of benefit to the crown.

The work of the Royal Corps of Engineers had wide scope, including the great military inspections of the Borderlands. Mapmaking, military preparedness reports, opinions on questions of geography, reports on economic matters and even general descriptions of the entire frontier are embraced within this great labor. These many documents written by early pioneer military engineers have been located, sifted and studied in a fruitful documentary recovery effort. In addition, Dr. Fireman has provided in her appendices and illustrations a broad sample of prototypes of this work, thereby giving the reader opportunity to appreciate more directly the range of labor undertaken. It is obvious in these reports that the engineers were really also geographers dealing with many aspects of an area's development, and that these documents are fine sources of information not regularly available to scholars.

Individual Royal Engineers, who up until now have been little more than names in an occasional listing of important personages in the Interior Provinces of New Spain, spring to life after nearly two centuries of oblivion. A happy combination of abilities permits the author to resurrect historically for the modern reader important individuals who were perhaps not even that widely known in their own lifetime. Dr. Fireman's knowledge of Spanish archives in Mexico and on the Iberian Peninsula, gained on the spot in months of research, and her comprehensive knowledge of the area of which she writes, give special vitality to her book. Her mastery of the language of the Borderlands military, not just the archaic Spanish encountered in neglected archives, but also her understanding of the specialized military jargon of the late colonial period, imbue authenticity with the words from her pen. This book brings to life a significant segment of the least-understood of the institutions of Spanish colonial control in the Borderlands – the military.

As contributions to learning, many books can be judged from the quality of an author's bibliography. Even a casual glance here demonstrates the preponderance of original manuscripts or contemporary published sources, while showing a scarcity of entries depending on secondary sources. This increasingly rare phenomenon, a book on a topic previously untreated, is both refreshing and rewarding to the reader. It is a characteristic of the Arthur H. Clark Company series dedicated to Spain in the West. Even the few familiar names such as Miguel Costansó and Nicolás de Lafora are offset by unknown and almost unpronounceable ones such as Juan de Pagazaur-tundúa.

In summary, the appropriate archives have been consulted, the winnowing process has been accomplished and the result is a book both masterfully compounded and skillfully written into a truly original historical study.

DONALD C. CUTTER

Preface

The Corps of Engineers in the Borderlands is not a common area of interest. Until now, nothing whatsoever has been published bearing directly on the subject. Research for my dissertation, and this book, then, was necessarily almost entirely primary in nature. The source materials were located almost exclusively in the archives of Spain and Mexico. Along with the requirement of distant research into a field of uncommon interest, the writer was blessed with an uncommon amount of invaluable aid and cordial advisers.

The guiding presence of an uncommon phenomenon, Professor Donald C. Cutter of the University of New Mexico, was the beginning, middle, and end of work on this project. It was through this extraordinary mentor that I made first contact with the Royal Engineers. While at the University of New Mexico in 1967, working on a master's thesis on presidios in the western Borderlands, I innocently asked Professor Cutter about this Corps, to which I had seen several vague references. No sooner put to the question than resolved! Professor Cutter demurely suggested that perhaps I had found an ideal dissertation topic. And because of him, perhaps I had, for his opportunely voiced suggestions, criticisms, and encouragement ghosted the dissertation to successful completion. It was his idea to appeal for the fellowship that the Woodrow Wilson Foundation so generously bestowed, which made journeys and work in Mexico and Spain during 1969 and 1970 so much more comfortable than they might have been without Wilson funding. It was also through his continuing support and interest in my work that the present publication arrangements were made.

Other uncommon helpers made the project complete. At the Bancroft Library in Berkeley, California materials were put at my disposal and photographic negatives from the fine Borderlands map collection at that institution were provided. The officers at the Archivo General Militar in the Alcázar at Segovia were polite and efficient as befitting the best traditions of the Spanish army. At the Archivo Histórico Nacional in the heart of Madrid, I was treated with unexcelled kindness. To the staffs of the great repositories for Borderlands research, the Archivo General de la Nación in Mexico City, and the Archivo General de Indias in Sevilla, great debts of gratitude are owed for the generous assistance they gave in tracking down totally uncatalogued items with only the barest clues as to their location. To the personnel at the Archivo General de Simancas in Valladolid, even greater thanks are offered. They acceded patiently to seemingly endless requests for more and more papers and bundles to be excavated from the dusty innards of the fantastic old castle where they are housed. Similar appreciation is felt towards the men and women at the Museo Naval in Madrid, who made archival research such a luxurious and pleasurable experience. Most of all, the officers and soldiers at the Biblioteca Central Militar, and its sister institution, the Servicio Geográfico del Ejército in Madrid, were inexplicably kind and accommodating to an alien researcher new to the world of engineers locked away for two hundred years in the archives. These worthy twentieth century successors to the Spanish army of the dissertation were inspiring, to say the least, in their enthusiasm for the project.

Further, interest and advice offered by Professor Manuel P. Servín of the Department of History at the University of New Mexico were irreplaceable tokens of faith that encouraged work step by step. Consuelo Boyd, of the Charles Trumbull Hayden Memorial Library at Arizona State University, graciously shared her excellent command of Spanish in completing and correcting the translations that appear as appendices to this book. Inexpressible gratitude goes to my father, Bert. M. Fireman, also of the Hayden Library and the Department of History at Arizona State University, to whom this book is fondly dedicated. I thank him for gently introducing me to the wonderful world of history at an early age; for ever more firmly encouraging my studies later on; for counsel and wisdom extended at every turn of my work; and for much more than simple paternal aid and concern in preparation of this book.

Additionally, classmates, colleagues, and friends spread across the western Borderlands, in Mexico and in Spain, where corpsmen originally labored, contributed helpfully to lighten the writer's labors in this attempt to introduce the accomplishments of Spain's Royal Corps of Engineers.

JANET R. FIREMAN
Los Angeles

The Spanish
Royal Corps of Engineers
in the
Western Borderlands

I

Organization of the Corps

Founded in 1711, the Royal Corps of Engineers was established in the wake of destruction and turmoil caused by the War of Spanish Succession. King Philip V of Spain encouraged foreigners, like the Marqués de Verboom who was father to the Corps, to help rebuild a war-torn, decadent empire. The Corps was born as an enlightened, well-organized institution of eighteenth century Europe.

Corpsmen were sent to the Indies in the first decades of the Corps' existence, but it was not until after the Seven Years' War in Europe that the Corps became truly significant on the northern frontier of Spain's American empire. The unfolding history of the Spanish Borderlands has been written largely in terms of the three great colonizing institutions employed throughout the vast region: mission, pueblo, and presidio. As an adjunct to the presidios and other military forces used by Spain on the frontier, the Corps performed invaluable, and as yet, uncalculated services.

Engineer Francisco Fersén participated in the important business of ascertaining the nature of the northwestern frontier in preparation for reorganization ordered by reformist Charles III and effected by Visitor General José de Gálvez. The Marqués de Rubí and his companion, Engineer Nicolás de Lafora, reconnoitered the Borderlands and proposed a new military organization for the Apache-troubled area.

Most famous among engineers was Miguel Costansó, who accompanied the Sacred Expedition that founded California in 1769, and then continued his engineering service in New Spain for forty-five years more. With Costansó, Engineer Alberto de Córdoba strengthened California's defenses, and Córdoba laid out the Villa de Branciforte, near present Santa Cruz.

Other engineers, like Gerónimo de la Rocha, Manuel Mascaró, Juan Pagazaurtundúa, and José Cortés, worked in the western Borderlands under the commandant general of the Interior Provinces. These men made important defense recommendations and improvements and executed many commissions of civil and church architecture as well as engineering projects like the construction of dams, roads, and bridges. They wrote descriptions of what they saw on the frontier that are invaluable historical sources today, and they drew maps that remain the best examples of eighteenth century cartographical knowledge of the area.

In the great variety of tasks that corpsmen performed, they helped create, enhance, and defend Spain's western Borderlands in that impossible dream of maintaining New Spain's northern frontier against the growing threats of foreign intrusion and internal dissidents.

In the history of the Spanish Borderlands, expansion northward from Mexico City has been viewed primarily through two of the three frontier institutions mentioned above: the mission and the presidio. Herbert Eugene Bolton set the stage and directed the players in the early years of Borderlands research. Starting with his work on missionary activities along the ever moving frontier,[1] his disciples have continued and expanded knowledge of Borderlands development. The presidio, or fort, the other great civilizing and expansion institution of the Borderlands, has been examined as representative of the defensive character of the Spanish northward thrust. Establishment of colonies for defense purposes, such as Texas and California; local organization of defense measures, as in the presidios and *casas fuertes* of the frontier; overall organization considered through military inspections and reorganization expeditions; special campaigns sent out for particular punitive objectives; and the establishment of military governments for the frontier provinces, have all been dealt with in pursuing the idea of settlement of the frontier as a defensive line moving steadily northward.

Missionaries, in their laborious and often inspired efforts at civilization and Christianization, moving from valley to valley up the western coast of Mexico, with all their zeal could not accomplish their objectives alone. The padres often entered new country with military escorts and time and time again, sol-

---

[1] See "The Mission as a Frontier Institution," Faculty Research Lecture, Univ. of California, Mar. 1917, first published in the *American Hist. Rev.,* vol. XXIII (Oct., 1917), pp. 42-61.

diers from the nearest presidios were called upon for aid. Even the frontier military establishment, itself never equal to the task, was at least partially dependent on outside assistance. The mere holding, much less extension, of the frontier was a huge responsibility. The Royal Corps of Engineers, a force which has been overlooked until this time, provided indispensable technical expertise to civilian and military officials for defense and extension of the frontier.

But the Corps of Engineers was not one of the basic institutions of the Borderlands. It was not active on every progressive frontier as were missionaries and soldiers. Not until the eighteenth century dawned, with the new ruling house of Bourbon in Spain, and the spread of awakened thought in that age, was there a specific, set organization for military engineers in the Spanish military establishment. The very nature of the Corps, technical, enlightened, specialized as it was, was both a product and a reflection of the times. Along with the many reforms they brought to government, society, and the economy, the Bourbon monarchs also revitalized the Spanish army. Part of resurrecting depleted Spain included rebuilding the army, where Bourbon reforms were especially significant. The first three Bourbon kings, Philip V, and his sons Ferdinand VI and Charles III, managed to create the first truly Spanish army since the days of Ferdinand and Isabella.[2] Among significant improvements and additions to the army was establishment of the Corps of Engineers.

[2] Antonio Domínguez Ortíz, *La sociedad española en el siglo* XVIII, p. 369. Unless otherwise noted, complete citations of reference works are given in the Bibliography.

The military reforms that led to creation of the Corps of Engineers began long before the peace was signed in Utrecht that ended the War of Spanish Succession. During this period of international rivalry and diplomatic intrigue, Spain underwent far-reaching reforms designed to increase its military strength and efficiency,[3] and which enabled the re-emerging nation to compete favorably in European and colonial affairs. Philip began to build a new army. While still containing elements of Spanish tradition, the new army would incorporate French and Prussian military innovations already proved effective on the battlefields of Europe. Mitigating against Philip's aims, certain aspects of the new European army were especially repugnant to traditional Spanish taste. Rigid order and discipline, absolute regimentation, and above all, obligatory service were offensive to Spaniards. But despite widespread popular objections to these reforms, new organizational patterns were necessary and requisite steps were taken. Through constant attention and adjustments, an effective armed force eventually was developed.

Creating the army required a tremendous output of administrative labor and material resources. In order to make expenditures of effort and money truly effective, public opinion toward the military required radical rehabilitation. In the preceding era, during the period of Spanish decadence, the people had come to hate the military, particularly in the absence of victories. During most of the seventeenth century,

---

[3] Charles E. Chapman, *A History of Spain,* p. 368.

only the most undesirable elements of society had chosen military careers; neither in battle nor in society was their comportment admirable. Yet, Philip V and his successors accomplished a complete transformation of public opinion as well as of the army itself through their reorganization efforts. By conferring economic, political, and social privileges on members of the military establishment, and by offering attractive inducements to the nobility, Philip made that group the foundation of a respectable officer corps. These honorable bribes bought success, for interest in the army as a profession among the aristocracy was quickly re-established.

Under the prior Habsburg conciliar system of administration, the Council of War had been the highest military authority. Under the new dynasty, the Bourbons asserted the prerogative, previously held by the council, for nomination to superior ranks within the military structure. The crown held power to nominate candidates for all employment in the army from lists submitted to the king by division or regimental inspectors. With this important power of appointment stripped from the council, it came to function primarily as an administrative agency meting out justice to members of the military services.[4]

The new Bourbon officer corps ranged from captains-general at the top, descending the ranks to second lieutenants. Internally, the army was organized in a modernized system of brigade, regiment, battalion, company, and squadron. These components were employed in four branches of the army, all of

---

[4] J. F. Bourgoing, *The Modern State of Spain,* vol. II, pp. 64-66.

which were created or stabilized by Bourbon monarchs and their ministers. Infantry, cavalry, artillery, and engineers were founded under the first Bourbon, and were destined to flourish under his sons.

On July 4, 1710, Don Jorge Próspero, the Marqués de Verboom, a Flemish nobleman, proposed a plan to the receptive king for establishment of a Corps of Engineers. By royal decree of April 11, 1711, the Corps was created and put into operation with Verboom in command as engineer general.[5] Organization and operation of the Corps began on a footing designed by Verboom that was to endure, albeit with numerous modifications, for more than a century. All engineers were required to have previous military training and status before they could be considered for the Corps. Once accepted, these officers encountered a rigid system of rank and class overseen by the engineer general. Below him were the stratified classes of director engineer, chief engineer, second class engineer, and ordinary engineer. At the same time the Corps was established, Verboom, the prime mover, organizer, and power of the Corps, established the Royal Military Academy of Mathematics in Barcelona in imitation of one in Brussels where he had studied.[6]

The infant Corps of Engineers, guided carefully by Verboom, began its long life actively and was to grow rapidly. On July 4, 1718, just eight years after the plan was presented to the king and only seven

---

[5] See Appendix A for a Spanish-English listing of rank and class in the Corps of Engineers from its founding to 1802.

[6] Special Compilation Commission, *Estudio Histórico del Cuerpo de Ingenieros del Ejército,* pp. 11-12.

years after the Corps' creation, a royal ordinance for governing the Corps was promulgated.[7] The explicit ordinance was divided into two separate sections: the first dealt with instructions for the formation of maps, plans, and drawings, and the second part concerned reconnaissance and review commissions. The plan also provided guidelines for calculations and estimates engineers were to make for construction, repairs, or enlargements of fortifications, barracks, royal installations, and other works and buildings. Classes and ranks were formalized and salaries and rations were specified.

The Ordinance of 1718 is notable in three ways. Excellent administrative and scientific principles were set down; extensive and far-reaching functions of the Corps were specified; and complicated and numerous duties were conferred on the seven year old Corps.[8]

In the following years, as the Corps of Engineers grew in size and stature, additional royal orders served to enhance the reputation of the organization. A royal decree in 1724, for example, prescribed an increase in salary and rations for the five classes of engineers, including the newly created class of extraordinary, or special engineer.[9] By 1728, just a year after the Marqués de Verboom died and his son took his place both as Marqués and as director general, the Corps numbered 127 men, including nine direc-

[7] José Antonio Portugués, ed., *Colección General de las Ordenanzas Militares, sus innovaciones, y aditamentos,* vol. VI, pp. 753-92.

[8] Special Commission, *Resumen Histórico del Arma,* p. 105.

[9] Portugués, *Colección General,* vol. VI, p. 796.

tors, nine chiefs, 27 second class engineers, 42 ordinaries, and forty extraordinaries.[10]

Changes in the organization of the Corps, and more precise delineation of rank and class took place after the younger Marqués de Verboom's death. A royal decree of October 19, 1756, declared that henceforth a chief engineer would have the rank of colonel of infantry, a second class engineer would be a lieutenant colonel, an ordinary would be a captain, an extraordinary would be a lieutenant, and a draftsman engineer would be a second lieutenant. The directors usually held the rank of brigadier or field marshal, and occasionally, lieutenant general.[11]

Also in 1756, a royal order issued by Ferdinand VI abolished the title of Captain General of Artillery and created the new title of Director General of Artillery, combining it in the same person as the Director General of Engineers. The Conde de Aranda was given this post,[12] but because of extreme confusion and inconvenience, the Engineers and Artillery were separated only two years later. Investigation of the fiasco, or union of the two corps, was again being considered in 1760. The king ordered Don Jaime Masones de Lima, at the French court, to find out all he could about the union of the two corps in France.[13]

10 *Estudio Histórico,* p. 12.

11 Portugués, *Colección General,* vol. VI, p. 803; and *Estudio Histórico,* p. 12.

12 "Royal Order of August 8, 1756," in Portugués, *Colección General,* vol. VI, p. 726.

13 The king to D. Jaime Masones, Jan. 19, 1760, Archivo General de Simancas, *Guerra Moderna,* 3802. Hereinafter cited as AGS.

Other delineations of the character of the Corps of Engineers took place while the Ordinance of 1718 was still in effect. The orderliness and efficiency for which the Corps strove was mirrored in various orders calling for progress reports on the work of each engineer. Two circulars of 1737 required provincial commanders to remit monthly lists of their engineers including the tasks to which each was assigned, and monthly notes on each engineer's activities. Again in 1747, a similar circular was distributed, requesting monthly lists of all engineers' assignments. Moreover, in 1762, the order was modified to require monthly lists of engineers and their assignments, and biannual reports on their conduct and progress.[14] This highly regulated and closely supervised system of operations was an outstanding earmark of Corps administration.

The king was not only interested in keeping close watch on his royal engineers; he was also concerned with their status within the military and in society. In 1763, the king, by this time zealous and well-loved Charles III, resolved that pension and retirement benefits be paid to officers of the Corps of Engineers in the amount of half of their salaries in their last employment. A royal order of July 31, 1768, reiterated and confirmed this decision.[15]

The special esteem in which royal engineers was held is demonstrated by a petition to the king and a royal decree issued in answer to it. Maximiliano de Croix, a member of the Corps and probably a mem-

[14] AGS, *Guerra Moderna*, 3002. Unfortunately, these reports for engineers in America have not been located in the archives.

[15] Biblioteca Central Militar, vol. 55, f. 93. Hereinafter cited as BCM.

ber of the illustrious Croix family that boasted a
viceroy of New Spain and the first commandant general of the Interior Provinces who was later viceroy
of Peru, wrote on behalf of the Corps of Engineers
to the Marqués de Squilache, Charles III's minister,
on July 2, 1765. Croix pointed out to the king that the
only manner of leaving the Corps at that time was
poor health or death. He suggested that in the case
an engineer was no longer able to carry out the rigorous activities associated with his profession, the
officer could still be of value elsewhere in the king's
service. Croix suggested that engineers be allowed to
retire from active service in the Corps, still retaining
their military rank, while serving in capacities such
as governorships, local commands, or district mag-
istrates. Croix explained that by employing engineers
in civil posts, the crown could save money, yet be
certain that difficult tasks would be well handled
because of the broad practical and theoretical training of engineers. Croix elaborated on this point,
saying that the majority of engineers were screened
from the most select cadets and officers of other
branches of the army, but that despite their application to the sciences and training in military theory,
they were usually put in positions of service under
military commanders with much less knowledge and
wisdom in the operation of military maneuvers. Croix
was afraid that without proper recognition and reward, incumbent engineers and those studying for
the Corps, would lose the inclination to continue their
careers. Further, he wrote, since engineers were
trained in reconnaissance, in mapping the routes for

marches and choosing bivouacs, in fortifications, in battle tactics, in siege, and blockade, in trench building and mining, in assaults and all other skills necessary for warfare and defense, the Corps of Engineers should have been regarded as a "true seminary for generals." Croix listed several engineers who had been placed in high positions of command in Italy and Africa, and cited the great job they had done. Croix concluded by saying that since engineers had shown themselves to be such good leaders during war, and such indefatigable workers in peace, there was no reason why they should not be given similar command posts in the peninsular provinces as well as in the colonies. The king answered Croix simply, yet powerfully, by announcing in a royal decree of August 20, 1765, that he would not now neglect, nor had he previously failed, in attending to his royal engineers.[16]

After a half-century of existence and increasing activity, the Corps had attained acknowledged status. Functioning as an adjunct to new Bourbon policies, the Corps enjoyed special prestige within military circles. Quality training, high standards, and sustained interest augured well for future developments that were to bring maturity and greatest glory to the Corps in Europe and in America. The first fifty years had pointed the way to participation by corpsmen in the momentous and eventful years to follow.

In accordance with the changing times and increasing complexity of operations of the Corps of Engineers, a new ordinance was handed down along with instructions for all other branches of the mil-

---

[16] AGS, *Guerra Moderna*, 2990.

itary. The new regulations, dated October 22, 1768,[17] were divided into three sections dealing with general regulations and engineers in garrison; fortification and other projects; and service of engineers in campaign. The Ordinance of 1768 had a distinctive character. Membership in the Corps was limited to 150 men, and operational procedures were so detailed that there remained few doubts as to procedures of the Corps and its relationship with military authorities. This precise document set the tone for operations of the Corps of Engineers in Spain and in the Americas for the remainder of the eighteenth century. Its importance as a constitution for corpsmen and as a historical reconstruction of their activities requires further examination.

According to the first section, from 1768 on, ranks and classes remained as they had been under the order of 1756, except that the class of draftsman engineer was changed in name to assistant engineer. The ranks were closed at forty assistants and forty extraordinaries; thirty ordinaries; twenty second class engineers; and ten each of chiefs and directors, all at the command of one engineer general. The 150 corpsmen were to be uniformed in the regulation army blue with a flesh-colored waistcoat, adorned coat with lapels and facings featuring silver buttonholes on one side and buttons on the other open flap. Each engineer's hat had a silver tape around it and matching silver galloons of two sizes on the shoulders of the coat. Matching rigs for horses were specified when an engineer was to be mounted.

---

[17] By royal order, *Ordenanza de S.M. para el servicio del Cuerpo de Ingenieros en Guarnicion, y Campaña.*

Before donning the elegant silver and blue of the Corps of Engineers, an applicant had to meet strict qualifications for acceptance. All men had to serve first as cadets or officers in the Infantry, Cavalry, Dragoons, Artillery, or the Navy. Petition for entrance had to be made to the engineer general, who would arrange for examinations in mathematics and drafting, subjects taught at the military academies at Barcelona, Orán, and Céuta. Having passed these examinations, and meeting other qualifications of age, health, and good conduct, applicants were reviewed by the engineer general when there were vacancies in the class of assistant engineer. Promotions in class were based only on length of service, and each man had to ascend each step of the rigidly ordered echelon. Promotions in rank for merit, or resignation and retirement from the Corps had to be specially petitioned through the king.

The Ordinance of 1768 spelled out specific functions and duties of the engineer general and the directors. Besides residing at court where he would be close at hand to the king, the engineer general was in charge of all top administrative functions and was directed to make tours of inspection of peninsular fortifications. Directors were assigned to a specific province where they had administrative and jurisdictional power over all the engineers stationed in the area. They assigned projects and duties to their subordinates and were responsible for all matters concerning engineers in the province.

For the first time, a special section on engineers assigned to the Indies was included in the Ordinance of 1768. The king was to approve appointments to America in all cases, and the special responsibility that these men had in the colonies was also reflected in their direct allegiance to the engineer general and his authority to issue special instructions to them. While in the Indies, an engineer's class position was considered vacated and therefore could be filled from lower ranks. As a supernumerary serving in America, an engineer still could be promoted in rank for longevity or merit, but when he returned to Spain, he would occupy the number in his class to which he belonged. If the class exceeded its prescribed number, no vacancies would be filled until the prescribed number again had been restored. Because of this restraint in promotion of engineers serving abroad, and because of the distance from home and hazards of journey and some assignments, service in the Indies was extremely unpopular. Cadets or officers in the Indies who served as volunteers for the Corps of Engineers were to petition for regular entrance only after passing the required examinations, and then with the recommendation of the viceroy, or commander general of the army.

Engineers assigned to the Indies were obliged to serve five years abroad. If, after those five years, they wished to return home to Spain, the engineer general had to provide replacements before the men could be released. Salaries in the Indies were to be the same as in Spain, with the addition of some special allowances for commissions that required lengthy and uncomfortable marches, similar to United States Army hazardous duty pay.

The second section of the Ordinance of 1768, dealing with fortifications projects and procedures, and
the third, concerning engineers' activities during
wartime, in siege, and in the field, are generally pertinent only to service in Europe and major colonial
harbor defense fortifications. Little in them applied
to New Spain at all, and almost nothing included
was pertinent to the western or northern Borderlands
frontier.

Only minor clarifications and additions to the regulations established by the Ordinance of 1768, were
made in the remaining years of the eighteenth century except for a major administrative change put
into effect by royal decree on September 12, 1774.
The Corps was divided into four independent sections, each with its own director. The first section
dealt with all sorts of military works and geographical projects; the second with civil buildings and
roads; a third with hydraulic projects; and the fourth
supplied teachers and administrators for the military
academies. Through trial and error, it was found that
the old, simpler system worked better and more
efficiently. The lack of one overall chief was sorely
felt. In 1797, therefore, the Corps of Engineers returned to the original, centralized system of organization, abolishing the four interior branches.[18]

In addition to the change of interior structure in
the Corps that operated from 1774 to 1797, the crown
acted to clarify certain points of the Corps' operation.

[18] *Estudio Histórico*, p. 27; *Resumen Histórico del Arma*, p. 110; and
Special Commission, *Compendio Histórico publicado al cumplirse el Segundo Centenario de la Creación del Cuerpo y Dedicado a sus clases e
individuos de tropa*, p. 34.

Because of the division of the Corps in its four sections, it was necessary to change the procedure for examination of applicants. A royal order of October 2, 1779, specified that applicants be examined at court by Don Pedro Lucuce, director of the branch in charge of academies.[19] Clarification was again needed a short time later when Lucuce died. Silvestre Abarca, director of fortifications, submitted certain proposals to the king concerning examinations. Abarca was worried about an apparent abundance of aspirants to the Corps; the fact that there were not enough vacancies to accommodate them; and that while waiting for entrance, without proper employment, the applicants forgot much of the technical knowledge learned in the academy. Abarca's worried letter and petition was passed on for comment to Don Francisco Sabatini, director of the branch dealing with roads, bridges, and civil buildings. Sabatini believed Abarca was over-anxious and concerned without reason, responding that the Corps was operating efficiently and effectively as it was, and that there were no abuses in the examination system, as Abarca had complained. Sabatini countered that the shortage of vacancies was not having the harmful effect that Abarca thought. Sabatini was confident potential corpsmen were not forgetting their training, and recommended that they continue to be allowed to wait, without returning to their regiments. The king apparently agreed with Sabatini that Abarca was worrying needlessly. On September 26, 1780, the king decreed that until a successor to

---

[19] BCM, vol. 55, f. 89.

Lucuce were chosen, all examinations would take place under the direction of Abarca and Sabatini together, with no changes being made in the established system.[20] The rigidly constituted system of examination for entrance to the Corps was therefore maintained and strengthened by the challenge of possible change. Reasoned tradition was a characteristic of the Corps by this time.

Only very occasionally were exceptions made in the explicit system of entrance to the Corps. In one instance in 1784, Don Juan Cavallero, a director and later engineer general, asked the king to grant his son, Manuel, the first vacant post among the ranks of assistant engineer. The king ordered this done, in recognition of the merits and service of Don Juan, and of his son. The latter was second lieutenant in the Infantry Regiment of Sevilla. He had participated in several battles and was wounded at Gibraltar. But the king admonished that this deviation was not to be taken as a precedent, but only as a special exception, since although Cavallero, the son, had been an officer longer than any other man in his class, he lacked necessary longevity as a volunteer engineer for promotion to assistant.[21] Obviously, the king endorsed the rigid promotion system articulated in earlier rulings, although he did make a single exception.

Equally exacting as the requirements and system for entrance in the Corps were standards of behavior that had to be maintained while serving as an engineer. To protect the elite position of the royal engineers, exceptional standards of work and discipline

---

[20] BCM, vol. 55, ff. 64-72.   [21] BCM, vol. 57, ff. 1437-38.

had to be maintained. Such was the essence of the royal order of May 16, 1787, stating that an engineer would be suspended from the Corps if he did not comply with orders or did not complete any assigned project. The order further stated that in promotions, particular talents and merits would be considered, not solely longevity in rank and class, especially when an engineer was available who seemed better prepared for promotion than the one next in line.[22] This change in royal policy for promotions can only be explained by an increased awareness of the importance of the Corps' work, bringing a new desire to inspire and reward exceptional service.

As in the Ordinance of 1768, particular attention in royal orders was directed to engineers serving in America. An order from the king dated May 6, 1787, was passed along to the Marqués de Sonora, José de Gálvez, who was serving as Minister of the Indies at the time. It was also sent to treasury officials and the governors and intendants of the frontier provinces by the viceroy, who was also archbishop of New Spain, Alonso Núñez de Haro y Peralta. The order had as its object the clarification of a longstanding problem regarding engineers in the Americas. Since the vastness of the countries necessitated long and inconvenient marches for engineers assigned there, a special stipend was to be awarded covering extra costs of these commissions. Engineers were to receive 25 pesos monthly in addition to regular salaries.[23] This order

---

[22] BCM, vol. 55, f. 91.

[23] Archivo General de la Nación, *Provincias Internas*, 121, and *Virreyes*, 142. Hereinafter cited as AGN. A peso in the eighteenth century had the approximate value of a United States dollar today.

recognized and corrected a problem which had been mentioned in the Ordinance of 1768, but at the time, the amount of extra compensation had not been designated, nor had exact requirements of eligibility for this extra pay.

Royal orders also answered questions involving engineers assigned to America and procedure for peninsular assignment upon their return. On September 27, 1787, the king required that in any case pertaining to the assignment of engineers in America, or any other matters dealing with engineers serving in America, questions and decisions should be referred to the Ministry of War, not to the Ministry of the Indies as had been done in several instances.[24] This order clearly strengthened the special character of the Corps' work in America, since ordinarily such matters would have been handled by the Minister of the Indies. When an engineer returned from his important and exhausting duties in America, he was to remain in the port of disembarkation until the director of the province received word of his arrival and thereupon gave the engineer notice of his new assignment in the peninsula. This was to save the engineer time and trouble in useless traveling.[25]

Possibly less important to the functioning of the Corps of Engineers, but treated just as judiciously and seriously, was a 1784 proposal by Sabatini to change the uniform of the Corps. Sabatini argued that both the dress and regular uniforms were overly elegant and expensive, and also, uncomfortable for normal use. He requested adoption of a simpler style,

---

[24] BCM, vol. 54, f. 472.
[25] BCM, vol. 55, f. 115.

similar to the artillery uniform. Taking note of Sabatini's criticism were royal orders of September 14 and October 25, 1784, approving his submitted designs, and providing that the dress uniform be worn only on feast days.[26]

Soon after the Corps' return in 1797 to its old and efficient organization operating under the Ordinance of 1768, an entirely new and completely different set of regulations was imposed upon it. The Ordinance of 1803, dated on July 11,[27] created the position of Generalissimo of the Corps, and gave that office to the Prince of Peace, Manuel Godoy, Charles IV's powerful and ambitious leading minister as well as Queen María Luisa's clever lover. Interior organization of the Corps was completely revamped, providing for 196 members arranged in a more rigid hierarchy. Slightly vague in places, the new Ordinance provided this new arrangement for engineers' service only in the peninsula, unreasonably leaving the former system to be observed in America. Norms for entrance into the Corps remained basically as they had been, but a new school for training was established at Alcalá de Henares, near Madrid. A modern innovation was establishment within the Corps of a regiment of sappers and miners. Probably the most interesting and salient change made by the Ordinance of 1803, was that the Corps of Engineers was reduced simply to another branch of the service, erasing former differences in nomenclature, privileges, and remuneration. Whereas under the Ordinances of 1718

---

[26] AGS, *Guerra Moderna*, 3002.

[27] *Ordenanza que S.M. manda observar en el servicio del Real Cuerpo de Ingenieros,* 2 vols.

and 1768, engineers were given complete liberty in arranging and drawing their maps, plans, and reports, the Ordinance of 1803 prescribed complete instructions and forms for every commission an engineer might be given.

The Corps of Engineers, then, operated from its very inception under a series of precise and well defined codes. It was the king's and his administrators' will that the Corps should in every way reflect the sophisticated principles upon which it was established, while executing its scientific and technical business. Though never composed of more than two hundred officers, the Corps of Engineers was integral to Spanish redevelopment and revitalization of the eighteenth century. It was an essential part of Bourbon reform in America as well as in the peninsula. Only a short time after its founding, far from home, in the northern reaches of Spain's American possessions, members of the Corps of Engineers were striving to live up to its lofty goals of enlightened achievement.

Arrival in the Borderlands

Within ten years of the Corps of Engineers' founding, a military engineer was hard at work on the northern frontier of New Spain providing an example of performance for other corpsmen who followed in later years. Francisco Álvarez Barreiro accompanied two military expeditions to the Borderlands in the decade between 1718 and 1728. Before that, little or no attention had been given to defense on the northern frontier. Owing to imperial considerations originating in Europe, the crown had been forced to concern itself primarily with fortifications and defense measures closer to the seat of the viceroyalty. Highest of priorities was harbor fortifications at San Juan de Ulúa, main port of entry for New Spain at Veracruz. Almost from the date of its founding, the Corps of Engineers took an active part in the maintenance and enlargement of that awesome citadel.

Other engineers preceded Álvarez Barreiro to New Spain. Sometime before August of 1722, in fact, an engineer was already resident in New Spain, probably at Veracruz, detailed for work on this major

fortification of the realm.[1] In 1722, the Marqués de
Castelar, one of the king's ministers, wrote to the
director of the Corps of Engineers, the Marqués de
Verboom, asking him to suggest a new man for the
post, because of the death of the first. The rigid
requirements for the job made it practically impossible for Verboom to find a qualified candidate. The
king wanted a Spaniard with knowledge and experience in fortifications near the water since almost all
the strong points in the realm were harbor posts.
Further, the candidate must be trustworthy, tech-
nically capable, and imaginative. He also should be
willing to go to New Spain.

Writing from Málaga two months later, Verboom
systematically went through the rolls of the Corps of
Engineers, discussing each man who might fit the
requirements for the job in New Spain.[2] Descending
from the higher classes of directors and chief engineers, Verboom finally endorsed an ordinary, the
Italian Felipe Leon Mafey, obviously not in the first
instance a Spaniard. He lacked experience in mari-
time fortifications, but had a good record in other
work in Sardinia. Verboom was uncertain about his
reaction to a trip to New Spain. Clearly, the young
Corps of Engineers did not yet have in 1722 a backlog
of experienced engineers on which to draw for important assignments of military construction and
maintenance.

Castelar responded to Verboom that after reviewing his last letter, the king had shown an interest in

---

[1] Marqués de Castelar to Marqués de Verboom, Madrid, Aug. 13, 1722,
AGS, *Guerra Moderna,* 2990.

[2] Verboom to Castelar, Málaga, Oct. 26, 1722, AGS, *Guerra Moderna,*
2990.

Mafey and two other men listed as candidates for the position. Verboom was thus directed to write to Mafey, and also to chief engineer Antonio Montaigu and second class engineer Andrés de los Cobos. He was to ask them if they would be interested in serving in New Spain, and to tell them that the trip would require nine or ten weeks, and would be financed for them and for their families by the crown.[3]

Less than three weeks later, such letters were sent to the three engineers in question.[4] Unfortunately, the resolution to this little drama is unknown. Responses to Verboom's questions by Mafey, Montaigu, and Cobos could not be located in the archives, and none of their names appear on lists of engineers in the Indies in the following years. But activities of the Corps of Engineers in New Spain were already under way. A permanent complement of engineers at Veracruz was established just a few years later.

By 1731, there was already one engineer, lieutenant and ordinary Fernando Gerónimo Pineda, stationed at Veracruz, and three more were proposed for the same post. The general rolls of the Corps for May of 1731 included three engineers with orders to go to Veracruz,[5] all of whom were to become notable members of the Corps and fit representatives of its highest ideals. Senior member of the trio was Captain Sebastián Feringnan Cortés, a second class engineer at the time, who was then stationed at Cartagena on

[3] Castelar to Verboom, Madrid, Dec. 11, 1722, AGS, *Guerra Moderna*, 2990.

[4] Verboom to Felipe Leon Mafey, Andrés de los Cobos, and Antonio Montaigu, Madrid, Dec. 30, 1722, AGS, *Guerra Moderna*, 2990.

[5] "Relacion gral de los Ings. que en el presente mes de mayo de 1731 están en actual servicio de S.M.," Barcelona, May 19, 1731, signed by the Marqués de Verboom, BCM, vol. 56.

the southeast coast of Spain. He had risen through the ranks quickly because of meritorious service. In recognition of his good record, as well as of his poor health, he was excused from assignment to New Spain at his own request in 1732. He continued to serve in the Corps at Cartagena, becoming chief of that defense arsenal. Feringnan retired from service on January 6, 1762, with the top rank of field marshal and director of engineers.[6] Joseph Reynaldo, a lieutenant and ordinary, was named in April of 1732 to replace Feringnan for the Veracruz assignment, but by July he had been ordered to Céuta instead, reaching there the following May from his previous post in Galicia.[7] Reynaldo apparently never went to Veracruz so Feringnan's position was left vacant.

But not so with the other two 1731 appointments to Veracruz. A brother of the engineer who was recently excused from new world duty, Felipe Feringnan Cortés, second lieutenant and special engineer at the time, left for New Spain from Cartagena where he had served beside his brother as inspector of sanitation. Don Felipe fulfilled a long and honorable career abroad, mostly at Veracruz, with a brief interlude at Pensacola. On August 23, 1766, he ascended simultaneously to the rank of colonel and to the class of director of engineers. The younger Feringnan died in service at Veracruz in 1769 with a long list of engineering credits in his base area.[8]

The fourth of the engineers appointed to the Vera-

---

[6] BCM, vol. 56.

[7] BCM, vol. 56.

[8] BCM, vol. 56; and José Antonio Calderón Quijano, "Ingenieros militares on Nueva España," *Anuario de Estudios Americanos*, vol. VI, 1949, pp. 47-49.

cruz mission in 1731 was to acquire a most distinguished record in America. Lieutenant and extraordinary engineer Luis Díez Navarro, a native of Málaga, who had been stationed in Cádiz, arrived in New Spain in company with Felipe Feringnan Cortés in 1732. For twenty years, Díez Navarro executed important military, civil, and religious architectural projects. His talents brought him into a close friendship with Viceroy and Archbishop Juan Antonio de Vizarrón. Unfortunately, this relationship also made him the rival and enemy of his direct superior, Italian-born Félix Prosperi. Prosperi was a fortifications expert who had arrived in Veracruz by 1737, and apparently believed military engineers should concern themselves exclusively with military matters. Prosperi's enmity notwithstanding, Díez Navarro carried out many baroque civil and religious commissions until about 1742, when he was transferred south to Guatemala as visitor of presidios in that realm. His name last appears on the rolls in January of 1780, then a brigadier and director of engineers.[9]

At the time of the 1731 assignments of Feringnan and Díez Navarro, which were to firmly establish the Corps of Engineers in the Veracruz and central valley areas of New Spain, another engineer had already returned from two long treks to the far north. Francisco Álvarez Barreiro set a precedent for engineers accompanying military expeditions. His first trip was in 1718-19 to Texas. Álvarez Barreiro had arrived in New Spain in 1716 with Viceroy Baltasar de Zúñiga, Marqués de Valero, and was designated by him as

---

[9] BCM, vol. 56; AGS, *Guerra Moderna*, 3794; Calderón Quijano, "Ingenieros militares," pp. 40-47 and 58-61.

official engineer for the Texas Alarcón expedition.[10]
The recently appointed governor of Coahuila, Don
Martín de Alarcón, was given the title of Captain-
General and Governor of Texas. He was assigned to
the frontier by the viceregal council to reinforce
Texas settlements established a few years earlier.
Continuing French interest in Louisiana after 1716
threatened the weakling Franciscans in Texas,
prompting Father Antonio de San Buenaventura
Olivares, a pioneer Texas priest, to travel to Mexico
and plead for reinforcements to establish a permanent
mission on the San Antonio River. Charged with
founding the mission for defensive as well as Chris-
tian purposes, Alarcón set off from the capital with
Álvarez Barreiro in the spring, and had arrived in
Texas by August of 1718. After establishing Mission
San Antonio Valero and the presidio of Béjar, Alarcón and Álvarez Barreiro traveled on, inspecting the
east Texas establishments. Various difficulties and
quarrels prompted Alarcón to resign his post and
withdraw from Texas.[11] Álvarez Barreiro returned to
the capital, and then to Spain, but his activities immediately following the Alarcón expedition are not
known to be documented.[12]

---

[10] Calderón Quijano, "Ingenieros militares," p. 20.

[11] John Francis Bannon, *The Spanish Borderlands Frontier, 1513-1821*,
pp. 116-19.

[12] Unfortunately, more complete information concerning the role Álvarez
Barreiro played in the Alarcón expedition has not yet been located. If the
engineer kept a diary or wrote a report, both likely possibilities, they re-
main undiscovered. The only known diaries of the expedition do not men-
tion Álvarez Barreiro. They have both been published as Francisco Céliz,
Diary of the Alarcón Expedition into Texas, 1718-1719, trans. and ed. by
Fritz Hoffman; and Fritz Hoffman, trans., "The Mesquía Diary of the
Alarcón Expedition into Texas, 1718," *Southwestern Hist. Qtly.*, vol. XLI,
Apr. 1938, pp. 312-23.

Better information is available concerning Álvarez Barreiro's duties and activities on his second trip to the northern frontier. This second trek was made under orders of Brigadier Pedro de Rivera, sent to review and inspect military garrisons. Previously, Rivera had held administrative posts in New Spain, among them the governorships of the presidio at Veracruz and of the province of Tlaxcala.[13] It is entirely possible that Rivera already knew Álvarez Barreiro and selected him specially as his technical adviser. The journey began on November 21, 1724, from the viceregal capital and ended 45 months and more than 8,000 miles later. Álvarez Barreiro, traveling with Rivera, first headed north from Mexico City, making garrison inspections along the way. By the spring of 1726, they had penetrated the central Borderlands, arriving in Chihuahua on April 7, 1726. After a tour north through El Paso del Norte to Santa Fé, they descended the Rio Grande as they had come, and crossed the Sierra Madre Occidental west of Janos, reaching Arispe on December 3, 1726. Leaving that future capital of the northern provinces, the Rivera party then headed south to Sinaloa, looped eastward back across the mountains through Janos, Casas Grandes, and Chihuahua; then went on to Texas, down to Monterrey and south back to Mexico City, entering the capital during the summer solstice of 1728.

Although Rivera's trip did not result in any tangible alterations of defense structure on the frontier, the expedition did serve several useful purposes. For

---

[13] Vito Alessio Robles, ed., *Diario y Derrotero de lo Caminado, Visto y Observado en la Visita que hizo a los Presidios de la Nueva España Septentrional el Brigadier Pedro de Rivera*, pp. 10-11.

the first time, an extensive frontier military review was accomplished, through which viceregal and home authorities were informed, also for the first time, on many matters of crucial concern. Moreover, the diary that Rivera himself kept listed a wealth of demographic and ethnographic data as well as distances and directions determined along his route. No doubt thanks to the technical training of Álvarez Barreiro, coordinates of no fewer than 29 settlements were determined for the first time. Besides filling a diary with concise and useful descriptions, Rivera offered recommendations that emerged from detailed reviews and hearings held at every presidio. Rivera was the first man to systematically and judiciously explore the defensive characteristics of the northern frontier. As a result of Rivera's trip, Viceroy Juan de Acuña, the Marqués de Casafuerte, handed down the 1729 regulations for presidios. These remained in effect until the more definitive and more appropriate *Reglamento* of 1772 was promulgated.

Unfortunately, neither diary nor documents written by Álvarez Barreiro have been found. But Rivera detailed in his diary various activities that the engineer undertook at Rivera's orders. Álvarez Barreiro was sent out from the main body of the party on four special mapping and reconnaissance expeditions. From February 10 to March 2, 1725, the engineer explored Nayarit and mapped the province with all its borders adjacent to Nueva Vizcaya and Nueva Galicia. Seven months later, Álvarez Barreiro left Rivera again, this time to map Nueva Vizcaya.[14] In

---

[14] Alessio Robles, *Diario y Derrotero,* pp. 35 and 38.

the following year, the engineer spent October 23 to November 21 charting the extensive domain of Sinaloa, Ostimuri, and Sonora. His final assignment was to reconnoiter and map Coahuila and Nuevo León, and after completing that during November of 1727, Álvarez Barreiro traveled to Monterrey where he met Rivera on February 24.[15] Although no diary or report of the Rivera expedition written by Álvarez Barreiro has been found, these four impressive maps, plus one of New Mexico, still survive in the archives of Spain. All are in color, on parchment, and are accompanied by a description of each of the territories.[16]

Little is known about Álvarez Barreiro after the important Rivera expedition. Apparently he left New Spain, having served his five-year assignment abroad. By royal decree of September 23, 1727, he was promoted to lieutenant colonel while still occupied in the north. Another decree dated July 15, 1729, confirmed the promotion and approved a proposed raise in salary. That order was probably issued upon the engineer's return to the peninsula.[17]

No evidence has been found indicating that the Corps of Engineers was active in the western Borderlands during the next forty years. In the first half of the eighteenth century, Spain went to war with England on four separate occasions. Imperial Britain's insults continued into the second half of the century

---

[15] Alessio Robles, *Diario y Derrotero*, pp. 57, 62, 83, 89.

[16] Archivo General de Indias, *Audiencia de Guadalajara*, 144. Hereinafter cited as AGI. Full titles, sizes, and descriptions of the maps are available in Pedro Torres Lanzas, *Relación Descriptiva de los Mapas, Planos &etc. de Mexico y Floridas existentes en el Archivo General de Indias, Sevilla.*

[17] Alessio Robles, *Diario y Derrotero*, p. 17.

however, and eventually forced Charles III, who had ascended the throne in 1759, to declare war once again. Spain signed the Family Compact with France and entered the Seven Years' War, a conflict which was to cause great changes in colonial America, where it was called the French and Indian War. The peace signed at Paris in 1763, after France and Spain had been beaten badly by English naval supremacy, was not regarded as decisive in Spain. Preparations were immediately begun for the next inevitable conflict with England. Terms in the 1763 treaty required France's virtual expulsion from the Americas. Spain gained sovereignty over all territory west of the Mississippi River, but had to cede her lands and claims east of the river to England. Other humiliating provisions of the peace and continuing English insults convinced Charles that radical measures had to be taken to prevent imperialistic England from absorbing Spain's overseas colonies as had already happened to France.

While the Spanish government concentrated on fighting European wars and colonial repercussions, the northern frontier of New Spain was neglected. Not until the turning point of 1763 did officials again consider the danger of foreign invasion through the Borderlands. With removal of France from the Mississippi area, Texas no longer was to serve as a defensive buffer against foreign intrusion. The primary danger was that England might attack New Spain by sea. To protect against this possibility, Charles and his ministers instituted a great number of economic reforms in the peninsula. Economy was the by-word for more than a decade, so that strengthening of Spain's military forces at home, in the colonies, and on the seas could be financed.

In addition to the fear of English attack by sea, officials in Mexico City were becoming increasingly aware of another formidable threat to the realm. This one, an interior menace, had to be considered at the same time that Spain was facing the foreign, or exterior threat. Hostile and rebellious Indians in the far western Borderlands were seriously endangering the safety of Spanish missions and settlements. Incessantly troublesome were the Apaches, who swept down on Spanish establishments in lightning raids, killing settlers and driving off large herds of livestock. The Pimas, generally docile and co-operative as mission charges, had gathered together in 1751 in a vengeful attempt to oust all Spaniards from their ancestral lands. Only desultory resistance was met by the Indians. Frontier presidios founded and located haphazardly, were in no condition to maintain an elementary state of defense, much less to help protect New Spain from potential British invasion.

Recognizing the various and complicated threats to his prime colony of New Spain, Charles III embarked upon a series of long-range projects designed to strengthen the western Borderlands, extend Spanish dominion and influence, and permeate the area with Spanish culture and tradition. In 1763, Charles did not know how far his reforms would reach. Nor was he aware of the repercussions triggered by his orders for military reorganization. Only a trifle of his many reforms is pertinent here, but that small bit is the basis for all of the activities of the Corps of Engineers in the Borderlands from then on. In the

Bourbon reforms, therefore, lay the foundation for the varied accomplishments of the royal engineers.

Beginning the chain of events that actually delivered the Corps of Engineers to the Borderlands was the military reorganization expedition sent to New Spain in 1764 under Lieutenant General Don Juan de Villalba y Angulo. In instructions dated August 1, 1764,[18] the king carefully spelled out for Villalba all that he was commissioned to accomplish: the establishment of defense works on Cortés' conquest route from Veracruz to the capital, and the complete reorganization of New Spain's military footing. The first of these projects required the work of the Corps of Engineers, and occupied many of its members for years to come. Detailed instructions explained to Villalba that from Veracruz, two roads branched to the west towards the capital. The more southerly passed through Orizaba about fifty miles from Veracruz, and the northerly road intersected Jalapa at about the same distance from the port. The two roads joined together at a place called Perote, located about 25 miles from both cities. It was at that junction that the king ordered Villalba to establish a fortress that would protect Mexico City in case the defenses at Veracruz crumbled during the expected sea invasion. The king further commanded that the roads themselves be made defensible. Many engineers throughout following decades were stationed at San Miguel de Perote, detailed for duty on the defense works there and on the roads leading to the fortress.

---

18 Instructions of the king to Villalba, San Ildefonso, Aug. 1, 1764, copy dated in Mexico, Jan. 21, 1765, AGN, *Indiferente de Guerra*, 304-A.

In addition to signaling the beginning of work on San Miguel de Perote, the Villalba expedition had further significance to the Corps of Engineers. Arriving in Veracruz on November 1, 1764, Villalba brought with him seven military engineers, including one lieutenant colonel, two captains, one lieutenant, and two second lieutenants.[19] Added to four members of the Corps already stationed in New Spain, Villalba's company brought the total to eleven. Of those men who arrived with Villalba in 1764 as part of Charles III's defense reorganization policy,[20] three traveled to the Borderlands in following years, where they made important contributions to the defense effort.

Although the king had invested Villalba with extremely broad powers for the military reorganization, he had instructed the military expert to co-operate fully with the viceroy. However, a complicated personal antagonism soon grew up between Villalba and the viceroy, the Marqués de Cruillas, which was partially to restrict efficient and effective completion of the projects ordered by the king.[21] Yet the personal problems between Cruillas and Villalba do not seem to have affected operations of the newly arrived engineers. A great deal of engineering activity began at this time, and continued throughout most of the colonial period.

A momentous event for the development of New

[19] AGI, *Audiencia de Mexico*, 2475.

[20] AGN, *Indiferente de Guerra*, 236.

[21] Lyle N. McAlister, "The Reorganization of the Army in New Spain, 1763-1766," *Hispanic Amer. Hist. Rev.*, XXXIII, 1953, pp. 8-18; and Vincent Peloso, "The Development and Functions of the Army in New Spain, 1760-1798" (unpublished M.A. thesis), pp. 46-47.

Spain was equally important for the future of the Royal Corps of Engineers in the western Borderlands. On July 18, 1765, Visitor General José de Gálvez arrived in Veracruz,[22] charged with a commission of reform and organization so vast that it was to mold the course of events in New Spain for years to come. As soon as the Visitor General arrived in the capital, about a month after setting foot on the shore at Veracruz, he presented himself to Viceroy Cruillas, and immediately afterwards to Villalba. Gálvez explained to the two men that the king had charged him with effecting a reconciliation between them so that royal plans for military reorganization and defense improvement might be continued without delay and complication.[23]

Gálvez could not have been completely successful in his efforts as peacemaker between the viceroy and the commandant general. In time, even Gálvez himself suffered friction in dealings with Cruillas. But his intervention certainly affected the Corps of Engineers. Its members were to participate in every major operation that Gálvez planned and supervised on the northern frontier, despite their being under the dual and divided orders of both Villalba and of Cruillas, who as viceroy, also was captain-general of New Spain.

The king's instructions to Gálvez, dated March 14, 1765,[24] charged him primarily with an inspection and reorganization of New Spain in which economy would be the motivating impulse. Gálvez was to

---

22 Joseph Ingram Priestley, *José de Gálvez, Visitor-General of New Spain, 1765-1771*, p. 136.

23 Priestley, *José de Gálvez*, p. 138.

24 Translated in Priestley, *José de Gálvez*, pp. 404-12.

oversee practically every aspect of administration in
New Spain, to accomplish reforms that would im-
prove the realm's fiscal and administrative policies.
Within Gálvez' sphere of activity, particular attention was directed to the northern border. Its importance in defense and as a potential source of new
revenues were of extreme concern to the crown. Initially, Gálvez was to bring peace to the Borderlands
in any manner he thought best. This region had been
suffering from Indian unrest for at least twenty years.
He was to examine the possibility and arrange for
establishment of semi-military settlements made up
of colonists recruited from the Mexico City area.
Mining, which had flourished in Sonora at one time,
was suffering from Indian assaults. Gálvez was instructed to promote and encourage mineral extraction
in every way he could. It would take a miracle to
accomplish. The number of presidios was to be reduced in the interest of economy, yet a more effective
military administration was to be established at the
same time. Because of the expulsion of the Jesuits in
1767, Gálvez was ordered to support new provisions
for the spiritual needs of the people in former Jesuit
domains, and was to participate in establishment of
a new bishopric. Further plans included the erection
of a sub-treasury at Álamos, in Sonora, for the collection of funds from silver mines on that remote
frontier.

All of these projected improvements for the northern areas were dependent on Gálvez' success in
achieving peace with the Indians.[25] Military engineers were active in the formal expedition that Gál-

---

[25] Priestley, *José de Gálvez*, p. 268.

vez sent north, as they were in almost every other project following the initial drive for peace with the Indians.

In September of 1765, Cruillas, Villalba, Gálvez, and others sat in military juntas that planned the expedition to Sonora. Gálvez named Colonel Domingo Elizondo to head the formal expedition. When combined with frontier troops, this force numbered about 1,100 men. Elizondo and 350 soldiers set out from Mexico to march to Tepic, near the Pacific coast, in April of 1767. There they waited until two brigantines were built at a new shipyard in San Blas. These carried them northward to Sonora, where they arrived in the harbor of Guaymas on March 10, 1768. That month operations against the Indians began and continued intermittently for almost three years. The Indians fled inland to retreats in the mountains of Cerro Prieto, and were unassailable there. The Spaniards eventually changed their tactics from frontal attack to guerrilla warfare, but generally they were unsuccessful. By the spring of 1769, therefore, Gálvez modified his former plans of expatriating or killing all rebel Indians. Instead, he offered them general amnesty. Shortly afterwards, Gálvez was taken ill with a fever, during which he suffered periods of insanity. During his illness, and while he was recovering in the spring of 1770, Elizondo continued with field operations in Sonora. By 1771, most of the Indians had submitted and were settled in small towns. Elizondo and his troops returned to Mexico.[26]

---

26 Even though Priestley and other modern historians have viewed the Elizondo expedition as something less than successful, official announcement at the time boasted success as seen in a revealing printed document

Two members of the Corps of Engineers were present during at least part of the Elizondo expedition. Both had come to America in 1764 with Villalba as second lieutenants and draftsman engineers. Since their arrival in New Spain, Miguel Costansó and Francisco Fersén had been stationed at Veracruz, and had worked there at various tasks of mapping and drafting.[27] Born in 1741 in Barcelona, Costansó had served in his native province, and then in Málaga after his acceptance into the Corps on January 12, 1762. In Veracruz, Costansó volunteered for the Sonora expedition under Elizondo and left the port in May of 1767.[28] He served with Elizondo in an unknown capacity until sometime in 1768, when he was transferred to the command of José de Gálvez. Costansó was briefly in Baja California, where he drew plans of the Bay of La Paz and Port of Cortés as well as projections for the Bay of San Bernabé and Cape San Lucas.[29]

The second engineer to join the Elizondo expedition was Francisco Fersén, who entered the Corps

published for general consumption. It was titled, "Noticia Breve de la Expedicion militar de Sonora y Cinaloa, su exito feliz, y ventajoso estado en que por consecuencia de ella se han puesto ambas provincias," Mexico, June 17, 1771, Archivo Histórico Nacional, *Diversos de Indias,* 464. Hereinafter cited as AHN. For more on the punitive campaign in Sonora, see Donald Rowland, "The Elizondo Expedition against the Indian Rebels of Sonora, 1765-1771" (unpublished Ph.D. dissertation).

[27] AGN, *Indiferente de Guerra,* 236.

[28] Service record, 1775, AGN, *Historia,* 568.

[29] "Puerto de la Paz sobre la Costa Oriental de California," Servicio Geográfico del Ejército, J-3-1-14. Hereinafter cited as SGE. The plan is reproduced in Michael E. Thurman, *The Naval Department of San Blas,* p. 64. Others of Costansó's Lower California maps are in AGI, *Audiencia de Guadalajara,* 416. The continuance of Costansó's career is taken up in Chapter IV. *See also* illustration on page 67, herein.

just eight months after Costansó, and was next on rolls for class promotions throughout their careers. Born in Paris, Fersén was 24 years old when he arrived in the new world. He had served as a volunteer in the Corps of Artillery in Portugal and in Madrid before receiving orders to report to Cádiz for embarkation with Villalba.[30] The exact nature of Fersén's responsibilities under Elizondo is as elusive for him as for Costansó. However, Fersén sat on at least one junta for formulation of attack plans during the campaign. In the fall of 1768, not long after the engineer's arrival in the north, Fersén, Elizondo, and the Sonora governor, Colonel Juan de Pineda, joined in conference at San José de Pimas, northeast of Guaymas. The three of them decided on assault plans against the rebel Indians. The maneuver consisted of two strong groups, under Elizondo and presidio Captain Bernardo de Urrea, marching simultaneously from Guaymas and Pitic (modern Hermosillo) and closing in on the Cerro Prieto in a pincer movement.[31]

While Fersén was in the north with Elizondo, possibly as part of his duties for the expedition and certainly with Gálvez' interests in mind, he wrote a short description of the area included in the governmental district of Culiacán, Sinaloa, and Sonora.[32] The Fersén document, dated in Pitic on January 2, 1770, deals primarily with the state of the mining

[30] Service record, 1777, AGS, *Guerra Moderna,* 3793; AGI, *Audiencia de Mexico,* 2424; and AGN, *Historia,* 568.

[31] Luis Navarro García, *Don José de Gálvez y la Comandancia General de las Provincias Internas del Norte de Nueva España,* p. 171; and AGI, *Audiencia de Guadalajara,* 416.

[32] "Descripción de las Provincias de Culiacan, Sinaloa y Sonora," Pitic, Jan. 2, 1770, BCM, ms. 5-3-9-14. See Appendix B.

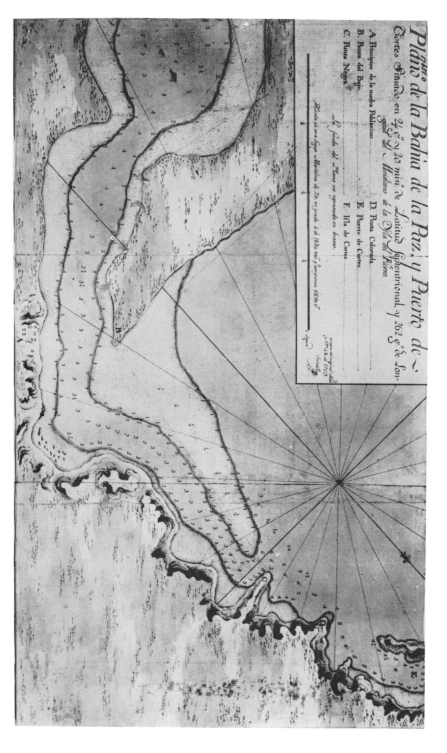

MIGUEL COSTANSÓ'S MAP OF THE BAY OF LA PAZ, 1769
See page 67. Courtesy, Bancroft Library.

community in the northwestern provinces and makes recommendations for economic improvements.

Fersén began his manuscript in the customary style used by engineers in the western Borderlands. He wrote a physical and topographical description of all the land he was speaking about, and from there discoursed on the mineral potential and prospective mining operations in the region. Generally, Fersén said, the provinces of Sinaloa and Sonora were rich and fertile. They produced grains and fruit as well as livestock of excellent quality. Fersén recognized the value of the Sonoran river valleys and lamented only that from the Yaqui River northward colonists were unable to make full use of nature's gifts because of marauding Apaches. In the jurisdiction of Sonora alone, north of the Yaqui, Fersén listed twenty mining towns or sites besides the major one located at San Antonio de la Huerta. Known also as Las Arenas, Fersén described the town on the west bank of the Yaqui River as a well laid out and pleasant settlement with a tranquil population. San Antonio surprisingly boasted about 25 dry goods stores and other retail firms as well, and supported trade from 400 to 500 pack mules carrying goods from Europe, Mexico, Puebla, and Guadalajara that arrived annually. Miners used San Antonio de la Huerta as their principal trading center according to Fersén, and farmers from the outlying areas came there to market agricultural products. Because miners came from far and wide to spend the yield of their labor in San Antonio, and because of its additional importance as a trading center for non-mining activities, Fersén thought it could serve as a royal revenue collection center. Fersén also suggested establishing Rosario as a sub-treasury for the jurisdictions of Culiacán, Sinaloa, and Ostimuri to the south.[33]

Fersén closed his report, written by order of the viceroy, Marqués de Croix, with a rhetorical question. He wondered what the future of the area would be if miners were free to pursue their work without the threat of Indian attacks, and with capital needs supplied by the government. One of many problems plaguing miners was that they suffered from a short-age of operating funds, so were unable to develop their undertakings to the greatest potential. As a result, many mining sites were left abandoned when the surface pickings had been reaped. Fersén insisted that with more concern invested in the area – that is, with the advent of governmental financial support for mining enterprises, and with increased military protection against hostile Indians – Sonora could not but prosper from its great mineral potential.[34]

After his 1770 report, Fersén apparently contrib-uted nothing more to the activities of the Corps of Engineers in the Borderlands. After serving in Sonora, he was recommended for promotion to cap-tain on September 7, 1771,[35] but that rank was not conferred until May 19, 1778, when he was made an

---

[33] Just six months after Fersén wrote his report, Álamos was established as a sub-treasury for Sonora. Priestley, *José de Gálvez,* p. 287.

[34] In his report, Fersén mentioned a map that he prepared to accompany his description. At the end of the document, the engineer added a note indicating that he had remitted the map and text together to the viceroy. Unfortunately, the map is not with the document in the Biblioteca Central Militar, and no other copy of it has been located.

[35] BCM, vol. 56.

ordinary engineer.[36] In 1782, Fersén traveled to
Havana to serve in the army of operations during the
war with England. On December 31, 1784, he was
awarded the rank of lieutenant colonel for meritorious service in Havana, and by December of 1789, he
had moved up to second class engineer. Fersén served
in Mexico City after his stint in Cuba until a royal
order of May 6, 1785, transferred him to Cartagena
de Indias.[37] After five years in South America and a
total of 26 years in the Indies, Fersén returned to the
peninsula and completed assignments in Navarre and
Valencia. Fersén became a colonel in 1793, and had
achieved the rank of brigadier and director of the
Corps of Engineers by the time of his death in 1805.[38]

But long before the completion of Fersén's illustrious years in Spain, other members of the Corps in
New Spain were implementing more of Gálvez' reorganization plans in the western Borderlands. At the
time when Costansó and Fersén were surveying the
Indian situation in Sonora with Elizondo, another
great expedition was shaping the future of the northern frontier. Led by the Marqués de Rubí and com-
plemented by a corpsman, its impact was great enough
to affect the present-day Borderlands area.

[36] Service record, 1787, AGS, *Guerra Moderna,* 5837.
[37] Service record, 1796, AGS, *Guerra Moderna,* 3794; and BCM, vol. 56.
[38] BCM, vol. 56.

3

Military Frontier Reorganization

Cayetano María Pignatelli Rubí Corbera y San Climent, the Marqués de Rubí, arrived in New Spain on November 1, 1764, with the Villalba military reorganization expedition. This much-respected nobleman served as Villalba's assistant for some ten months before being commissioned by the king to carry out an extensive military inspection of the northern frontier of New Spain, as Rivera had done forty years earlier.[1]

Gálvez, with the king's orders in hand for executing widespread reforms in New Spain, reached Mexico just a month before Rubí received his commission as military inspector of the Borderlands. The missions of these two men were entirely separate in theory, although related in practice. While Gálvez was entrusted primarily with financial and related administrative reforms, he chose a military project in Sonora as the first necessary objective to his reform process. Rubí, on the other hand, was to deal exclusively with the frontier defense situation. His recommendations and their subsequent implementation

---

[1] AGI, *Audiencia de Mexico,* 2475; and royal order of Aug. 7, 1765, AGI, *Audiencia de Guadalajara,* 511.

remodeled the frontier organization and coincidentally affected the future delineation of a Mexico-United States border.

The magnitude of Rubí's commission was immediately evident. He was detailed to inspect every frontier presidio as thoroughly as possible and to make recommendations for improvement of northern defense. Consequently, he required the expertise of a trained engineer. Viceroy Joaquín de Montserrat y Ciruana, the Marqués de Cruillas, appointed Nicolás de Lafora to serve as Rubí's technical right arm [2] and instructed Rubí that when sending reports back to the capital on each presidio visited, he should include a small map or sketch drawing by the engineer, "in order to show better the country in question." [3]

Lafora was an excellent choice for the assignment. By the time of his appointment to the Rubí inspection, he had been in the army almost twenty years, and had an impressive record. Born in Alicante, Lafora began his career as a volunteer in the navy for eleven months and then transferred to the army as a cadet in the Infantry Regiment of Galicia where he served for more than ten years, the last two as ensign. He fought in Italy, Africa, and Portugal in at least six major campaigns, two sieges, assaults on two fortresses, and six field battles.[4] From July 1, 1752, until the last day of June 1755, Lafora was a student in the Royal Academy of Mathematics in Barcelona. His certificate of graduation from the Academy indicates that he finished the course of studies with

---

[2] Lawrence Kinnaird, trans. and ed., *The Frontiers of New Spain: Nicolas de Lafora's Description, 1766-1768,* p. 44.

[3] Instructions of Marqués de Cruillas to Marqués de Rubí, Mexico, Mar. 10, 1766, AGI, *Audiencia de Guadalajara,* 273.

[4] Service record, 1769, BCM, vol. 57.

distinguished accomplishment in defense theory and drafting, the two principal subjects of the curriculum. It also reveals that Lafora excelled in special skills, application to work, and punctual service.[5]

In October of 1756, fourteen months after finishing academic studies, Lafora applied for admission to the Corps of Engineers. His name was proposed for acceptance on December 27 of the same year, and Lafora officially entered the Corps on January 15, 1757, as second lieutenant and draftsman engineer.[6] Immediately assigned to Barcelona, he served there until 1760. During the next two years he worked in Aragon and then was transferred farther west to Galicia and then back to Catalonia in April of 1763. Like Costansó and Fersén, Lafora came to New Spain as part of the Villalba expedition, but he had been designated only at the last minute and as a substitute. In a royal order of June 1764, Lafora was named to replace engineer Bernardo Lecoq in New Spain, while Lecoq remained in Catalonia. Lafora left at once from Barcelona for embarkation at Cartagena.[7]

Before going to the Indies, Lafora's record had been outstanding. Throughout his career, petitions and letters to his superiors show his great enthusiasm, and even greater ambition. Lafora was proud of his work, anxious to serve. There was only one blot on his record. Causes of the scuffle are unknown, but in November of 1757, Lafora was arrested for beating two townsmen of Figueras with a sword. The newly appointed engineer was arrested and imprisoned for

---

[5] Archivo General Militar, *Expediente personal de D. Nicolás de Lafora.* Hereinafter cited as AGM.

[6] AGM, *Expediente personal de D. Nicolás de Lafora;* and BCM, vol. 56.

[7] BCM, vol. 56.

a short time.[8] Although Lafora is not known to have been imprisoned again, he had the courage – or audacity – to disagree with his superiors in later years.

By January of 1762, Lafora was already an extraordinary engineer, and on March 19 of the following year he became captain and ordinary. Soon after arrival in New Spain in 1764, Lafora was assigned by Cruillas to accompany Rubí on the northern inspection. Lafora left a diary of the Rubí expedition that has been called the finest available source of information on the northern frontier of New Spain.[9] Lafora carefully included descriptions of geographical features, population statistics, frontier conditions, and suggestions for defense improvements. Lafora's daily account of travels, covering some 7,600 miles and lasting 23 months, gave detailed information of presidio reviews and inspections in the provinces of Nueva Galicia, Nueva Vizcaya, New Mexico, Sonora, Coahuila, Texas, and Nayarit.

Although Rubí was appointed directly by the king and was responsible only to him through Julián de Arriaga, Minister of the Indies, instructions were given him by the viceroy. There was some delay in Mexico awaiting delivery of Cruillas' instructions, but upon receipt of them on March 10, 1766, Rubí set out for the north.[10] The emphasis of Cruillas'

---

8 AGM, *Expediente personal de D. Nicolás de Lafora.* Figueras is probably the town known today by the name of Figuera, northwest of Barcelona about fifty miles. The tussle took place on Nov. 21, and the notice of it and Lafora's arrest is dated in Barcelona, Nov. 26.

9 Kinnaird, *The Frontiers of New Spain,* p. 3.

10 See Navarro García, *Don José de Gálvez,* p. 136, n. 9; and Kinnaird, *The Frontiers of New Spain,* pp. 3-4, for a discussion of the delay and disagreement between Cruillas and Rubí. For some reason, Lafora left Mexico a week after Rubí, on Mar. 18, and caught up with him on Apr. 14 at Durango. Kinnaird, *The Frontiers of New Spain,* p. 59.

instruction was on Sonora and was accompanied by a number of helpful documents and maps. These included a report from the Auditor of War setting forth complete information on every presidio; a map of the entire northern provinces, showing all presidios; a more detailed map of just Sonora and Sinaloa; and the map made by Francisco Álvarez Barreiro, who accompanied Pedro de Rivera forty years earlier.

On April 30, 1766, Rubí and Lafora reached the presidio of El Pasaje, about eighty miles northeast of Durango, where they began their first garrison inspection. While Rubí carried out his assigned duties of examining men, mounts, records, and equipment of the fort for two weeks, Lafora wandered over the area making notations and taking astronomical observations. This procedure was to be repeated time and time again as Rubí and Lafora marched back and forth across the Borderlands in following months. From El Pasaje, the party continued north to Chihuahua, through El Paso, and up the Rio Grande valley to Santa Fé. Then they turned back on their tracks, as Rivera and Álvarez Barreiro had done, descending the river and crossing the mountains to Janos and on to the six Sonora presidios. Rubí and Lafora spent Christmas and New Year's day of 1767 at the presidio of Tubac where Captain Juan Bautista de Anza the Younger was their host. After reviewing Sonora, Rubí and Lafora crossed the mountains eastward to Coahuila, Texas, Nuevo León, and then turned south again to the viceregal capital, arriving there on February 23, 1768.

Back in Mexico City, Lafora worked on the final drafts of his report and maps while Rubí composed his *dictamen*.[11] Rubí made suggestions to the crown which included his innovative plan for a cordon of fifteen presidios stretching across the northern border of New Spain. He realized what others had over-looked too long: Spain had spread herself too thin over too great a territory. The sensible measure to be taken, according to Rubí, was to delineate her true domain, and protect that from further aggression. It was this astute awareness which Herbert Eugene Bolton singled out as Rubí's greatest contribution to northern defense.[12]

The realignment of the presidial cordon suggested by Rubí, and most of his other defense suggestions, were approved and put into effect by a royal order of July 11, 1769.[13] A direct result of Rubí's and Lafora's inspection tour was the issuing of the Regulations of 1772,[14] calling for the creation of the Interior Provinces and a whole new defense organization. The new defense ideas eventually evolved into the Commandancy General of the Interior Provinces, in which military engineers played a prominent role.

Lafora is his own chronicler for the Rubí inspection tour.[15] The thoroughness of his descriptions, the

11 "Dictamen que de orden del Excmo. Señor marqués de Croix, virrey de este reino, expone el mariscal de campo marqués de Rubí en orden a la mejor situación de los presidios para la defensa y extensión de su frontera a la gentilidad en los confines del norte de este virreinato," Tacubaya, Apr. 10, 1768, AGI, *Audiencia de Guadalajara,* 273 and 511; also, *Audiencia de Mexico,* 2422 and 2477.

12 Herbert Eugene Bolton, *Texas in the Middle Eighteenth Century,* p. 379.

13 AGI, *Audiencia de Guadalajara,* 511.

14 Facsimile reproduction of the 1834 Mexican edition and translation of this *Reglamento* are published in Sidney B. Brinckerhoff and Odie B. Faulk, *Lancers for the King,* pp. 12-67.

flow of his prose, and the accuracy of his astronomical observations, all demonstrate technical expertise. Without a doubt, Lafora was of great help to Rubí,[16] and the lasting importance of their joint reorganization plan for the frontier is reflected in the implementation steps taken by other frontier administrators.

Existing copies of the Lafora diary reveal the prominence of his work and the esteem in which military engineers were held. No fewer than seven partial and complete copies of the diary have been located in the archives of Spain, Mexico, and the United States.[17] The number of these, and the dates

[15] Lafora's full diary has been published in both Spanish and English. The Spanish edition is edited by Vito Alessio Robles, *Nicolás de Lafora: Relación del viaje que hizo a los Presidios Internos situados en la frontera de la América Septentrional perteneciente al Rey de España.* The English edition, already cited, is Kinnaird, *The Frontiers of New Spain.*

[16] Rubí said that Lafora's diary, observations, and maps, as well as his counsel, were "of the greatest utility" to him in the formation of his *dictamen.* Rubí to Juan Gregorio Muniain, Barcelona, Mar. 31, 1770, AGS, *Guerra Moderna, 3089.*

[17] Kinnaird lists extant copies and fragments of Lafora's diary in the Biblioteca Nacional in Madrid (ms. 5963), the Biblioteca Nacional in Mexico City (ms. XV-2-60), and the Bancroft Library in Berkeley. Navarro García notes two more copies of the diary, both unsigned, in the Biblioteca Central Militar. They are entitled, "Situacion en que se hallan todas las provincias del reino de Nueva España fronterizas a la gentilidad en las partes del norte" (ms. 5-3-9-5), and "Diario del reconocimiento de una parte de la América septentrional española, 1776" (ms. 5-3-9-8). Appended to the second copy of Lafora's diary by the BCM is the instruction for and journal of a trip made by Juan María de Rivera covering the region from Texas to Nayarit in 1765. Rivera signed himself as an engineer but was not a member of the Corps at any time. Like others, he could have been an engineer attached to an infantry or cavalry grouping. Another copy of Lafora's diary, not mentioned in any published sources, exists in the Museo Naval, ms. 334 (Reino de México II). Hereinafter cited as MN. It is unsigned, but a complete copy. Also in MN, ms. 567 (Virreinato de Mexico I), there is an unsigned, undated document entitled, "Informes sobre las provincias internas por el ingeniero Lafora." This copy was made by order of the viceroy in 1777.

that the copies were made indicate that Lafora's information was still considered valuable many years after his trip to the Borderlands.

Besides the diary, Lafora left several other documents from the Rubí expedition. Like Rubí, Lafora wrote a *dictamen,* a document containing suggestions and opinions. The engineer's *dictamen* was an extensive report on the defensive problems of Nueva Vizcaya and included recommendations for improvement of the military situation.[18] Besides the small sketch maps that Lafora must have drawn to accompany periodic reports to the viceroy according to instructions, Lafora drew an excellent surviving map, considered the best available for years to come. Based on his own observations of latitude, Lafora's map shows the route traveled by the engineer and Rubí in the frontier provinces. It covers an area from Papagoland near the junction of the Gila and Colorado Rivers in the Pacific west to the Sabine River at the present Texas-Louisiana border in the Gulf south. Taos, New Mexico is the most northerly point shown on the map, which extends southerly as far as Nayarit in the west and Tampico in the east. Both the extant locations of presidios and suggested new locations on the frontier line are shown by Lafora. Additionally, Lafora included cities, villas, and towns; missions, *reales de minas,* and Indian settlements; ranches and haciendas; and abandoned settlements of various types. Rivers, roads, and trails are delineated clearly.

18 "Dictamen que para asegurar las fronteras de la Nueva Vizcaya da el capitán de Ingenieros Don Nicolás de Lafora. Fundado en lo que ha visto de ellas, en los informes de la gente más práctica y en los mapas más correctos de este pais," Chihuahua, July 2, 1766, AGI, *Audiencia de Guadalajara,* 511. *See* illustration on page 85, herein.

Stylized mountains make it somewhat difficult to determine relative changes in terrain, but Indian regions and place names are spelled out clearly in Lafora's round, even script.[19]

Other valuable maps, often thought to be the work of the Corps of Engineers, came out of the Rubí expedition. The name of Don Joseph de Urrutia is well known among Borderlands cartography *aficionados* for the series of maps that he drew as part of the Rubí inspection. Actually, Urrutia was never a member of the Corps, but had engineering training. He served as second lieutenant and draftsman of the Infantry Regiment of America, and was assigned for special duty to accompany the review tour.[20] Urrutia traveled the frontiers with Rubí and Lafora, and obviously worked closely with the army engineer. Urrutia recorded plans of more than twenty presidios and settlements visited and inspected by Rubí, as well as plans of the attack on Cerro Prieto staged by Elizondo and his men.[21]

---

[19] The original of Lafora's map, dated in Mexico on July 27, 1771, is in SGE, J-2-3-96, and there is a contemporary copy in the British Museum Manuscript Room, Add. 17660a. Another copy is located in the Secretaría de Fomento in Mexico, Sección de Cartografía. There is a photograph of the original reproduced in Navarro García, *Don José de Gálvez*, following p. 176; and the map was redrawn in a legible manner, imitating Lafora's style and script by Inez Durnford Haase, included at the end of Kinnaird, *The Frontiers of New Spain*. A similar reproduction is attached to the back cover of Alessio Robles, *Nicolás de Lafora*. Lafora's map is also reproduced in the extremely helpful collection of maps of the new world, *Cartografía de Ultramar*, vol. II, no. 122. *See* page 85, herein.

[20] AGI, *Audiencia de Mexico*, 2459; and AHN, *Estado*, 3882.

[21] Urrutia's presidio plans are in the British Mus. Ms. Room, Add. 17662 a-x, and are described and reproduced in Navarro García, *Don José de Gálvez*, following p. 176, and pp. 536-39. Urrutia's plan of the attack on Cerro Prieto is in AGI, *Audiencia de Mexico*, 2430; and a contemporary copy of it by D. Luis Surville is in the British Mus. Ms. Room, Add. 17651p.

Lafora too, drew presidio plans. Although that was Urrutia's job primarily, Lafora produced maps of the central Texas presidio of San Sabá and also of Guajoquilla on the Conchos downriver from El Paso. Both of them were designed to show the defensive disadvantages of the sites as part of Rubí's and Lafora's recommendations that those two garrisons be relocated.[22]

Even before returning to Mexico at the end of the two-year inspection tour, Lafora embarked on the next phase of his life's work. At a time that should have been the apex of his career and ambitions, Lafora began a twenty-year struggle that led only to frustration and disappointment. Writing from the presidio of San Miguel de Horcasitas, Sonora, in February of 1767, Lafora asked for promotion to the rank of lieutenant colonel. As was the custom, he wrote copiously, describing the action he had seen and services he had rendered to the crown while in the army. Lafora was a captain at the time, and had held that rank for only four and a half years. Don Juan Martín Zermeño, director general of the Corps, gave his judgment on Lafora's petition. Zermeño said that because of Lafora's short service as captain, because there were thirty other engineers in his class with greater longevity; and because he had not performed any task above and beyond the call of duty, Lafora was in no way considered deserving of promotion to lieutenant colonel.[23]

---

[22] Plan of Guajoquilla, AGI, *Audiencia de Mexico,* 578; plan of San Sabá, AGI, *Audiencia de Mexico,* 579. Both are described completely in Torres Lanzas, *Relación Descriptiva,* and are reproduced in Navarro García, *Don José de Gálvez,* following p. 176.

[23] Lafora to Arriaga, Horcasitas, Feb. 21, 1767; and Zermeño to Muniain, Barcelona, Oct. 28, 1767, AGS, *Guerra Moderna,* 3089.

Lafora's presumptuous request and its subsequent prompt denial did not discourage the ambitious engineer. Just short of two years later, he wrote to Zermeño reporting on the Elizondo campaign in Sonora, and pointing out that his and Rubí's long-range plans for the defense of the frontier would undoubtedly be more effective than costly punitive expeditions such as Elizondo's. In the letter, Lafora also took the opportunity to inform Zermeño that he had again petitioned for promotion to lieutenant colonel, this time with strong recommendation from the Marqués de Rubí. Lafora expressed the hope that he be conceded the promotion and a governorship of some province in America, since the fatigues of the Borderlands inspection had broken his health.[24]

A third time, in May of 1769, Lafora petitioned for promotion to lieutenant colonel. Now he wrote to the king, including a copy of his service record. Lafora was by no means shy in reciting his services to the crown. He elaborated on the dangers and discomforts of his journey with Rubí, and patted himself on the back for having drawn a map that was "the most accurate that there has been to this day." In Spain, Lafora's petition was passed to the Marqués de Rubí for comment. The engineer's former commander and companion certified that everything said in the document was true and that he considered Lafora worthy of the promotion.[25] Nevertheless, Lafora's request again was denied.

During the period between Lafora's first and sec-

---

[24] Lafora to Zermeño, Mexico, Dec. 5, 1768, AGM, *Expediente personal de D. Nicolás de Lafora.*

[25] Lafora to the king, Mexico, May 26, 1769, BCM, vol. 57; and Rubí to Muniain, Barcelona, Mar. 31, 1770, AGS, *Guerra Moderna, 3089.*

ond requests for promotion, he returned to Mexico
City upon completion of the inspection tour. Arriving in Mexico in February of 1768, Lafora worked
on drainage projects for the capital when he was not
occupied with petitioning his superiors or singing his
own praises. He directed various works for containing Lake Texcoco which threatened to inundate the
barrio of Santa Cruz, near the capital.[26] Meanwhile
Lafora kept busy with events other than just his engineering assignments. Just three months after his return from the north, when he was almost forty years
old, the engineer requested and was granted license
from Zermeño to marry Doña Francisca de Arce y
Echegaray.[27]

After still another petition by Lafora in 1770,
Zermeño still refused to promote Lafora, saying that
he did not yet deserve advancement in rank, but that
he did deserve to be relieved of duty in America, having served the mandatory five years.[28] Royal orders
that same year relieved Lafora in America and
ordered engineer Francisco Calderín, serving in
Cuba, to replace him in Mexico. Before leaving New
Spain, Lafora petitioned Zermeño once again for a
favor.[29] By the middle of 1772, Lafora had returned

[26] AGN, *Indiferente de Guerra,* 331; and Alessio Robles, *Nicolás de Lafora,* p. 18.

[27] AGM, *Expediente personal de D. Nicolás de Lafora.* Members of the Corps of Engineers were required to assemble copious documentation on the families of prospective wives and request formal permission and government license before marrying.

[28] Zermeño to Muniain, Barcelona, Aug. 4, 1770, AGS, *Guerra Moderna,* 3089.

[29] Royal order of Aug. 16, 1770, San Ildefonso, AGS, *Guerra Moderna,* 3089; and Lafora to Zermeño, Mexico, Feb. 28, 1771, AGM, *Expediente personal de D. Nicolás de Lafora.*
Included with other documents concerning Lafora's requests for promo-
tion between 1767 and 1770, is a petition signed by Lafora and also by his

to Spain, where he was assigned to serve in Galicia. He requested a change to his native Alicante, which was conceded in April of 1773.[30]

During this period when Lafora was jockeying for advancement and directing projects in Mexico City, he was recognized as an expert on the Borderlands. From time to time he wrote to officials in Spain about events in the north, and prepared a map of the southern coast of California.[31] Shortly after his return to Spain, in 1772, he was called to a junta held at the home of the Marqués de Croix, former viceroy of New Spain. Lafora reviewed for the assembled officials recommendations that he and Rubí had made concerning the defense of the Borderlands.[32]

For the next few years, Lafora served in Alicante as commandant of harbor fortifications.[33] In 1774, in answer to Lafora's longstanding request, a royal order appointed the engineer *corregidor,* or district magistrate, of Oaxaca – back in New Spain. On April 18, 1775, the engineer turned civil servant embarked from Cádiz for New Spain, where he was to take up his post in Antequera.[34] The appointment was

brother. At the end of the usual form request, the document reads: "En virtud de poder de mi hermano Dn Nicolas Laffora, Pedro Antonio Laffora (rubric)." Don Nicolás almost always spelled his family name with one "f." His brother's seconding of his application for promotion is an unusual phenomenon, as he was not a member of the Corps, and apparently had nothing more to do with the matter than brotherly concern.

30 Royal orders of Apr. 18, 1772, and Apr. 8, 1773, BCM, vol. 56.

31 "Plano de la Costa del Sur dela California" in "Catálogo de los documentos y planos descriptivos del Reino de Nueva España, 1824," BCM, ms. 5-2-4-12. This map no longer exists in the archives of the BCM or SGE.

32 Alessio Robles, *Nicolás de Lafora,* pp. 18-19.

33 AGN, *Indiferente de Guerra,* 331.

34 AGM, *Expediente personal de D. Nicolás de Lafora.* Lafora's appointment to this assignment is an example of Maximiliano Croix's 1765 suggestion put into operation. See above, pp. 36-38.

for the usual five years, but because of the sudden death of the man appointed to replace him at that time, Lafora continued to serve in Oaxaca an extra three years. During that extended period, he finally ascended to the rank of lieutenant colonel in the army, the honor to which he had aspired for at least ten years. As chief administrator of Oaxaca, Lafora was honorable, diligent, and competent. Besides his civic duties, he erected city buildings in Antequera that Lafora himself thought to be among the most beautiful in all of America.[35]

In 1785, Lafora was relieved of the *corregimiento* of Oaxaca. He was left without employment and without income, and was forced to support himself through farming. Lafora asserted that his health was broken, and he was forced to ask for a loan to maintain his meagre agricultural endeavors.[36] In January of 1786, José de Gálvez, then Minister of the Indies, passed on a petition to officials of the Corps of Engineers in which Lafora asked for promotion to the rank of colonel with full retirement benefits. Corps officials denied judgment on Lafora's case, saying that since he had left the Corps for political employment, they no longer held jurisdiction over him for retirement benefits or any other considerations.[37] The Corps left the matter in the hands of Gálvez.

---

[35] Description of the 1781 plans for the facade of the Antequera city hall drawn by Lafora can be found in Torres Lanzas, *Relación Descriptiva,* vol. II, p. 43.

[36] Herbert Eugene Bolton, *Guide to Materials for the History of the United States in the Principal Archives of Mexico,* p. 220; and Alessio Robles, *Nicolás de Lafora,* p. 24.

[37] Marqués de Sonora (José de Gálvez) to Inspectors of the Corps of Engineers (Sabatini and Cavallero), the Palace, Jan. 3, 1786; and Sabatini and Cavallero to the Marqués de Sonora, Madrid, Jan. 16, 1786, AGM, *Expediente personal de D. Nicolás de Lafora.*

Gálvez must have decided to deny Lafora promotional retirement with pension, because in 1788, the frustrated Lafora made one final, futile attempt to petition the king. He requested return to the Corps of Engineers with the rank and class he would have attained had he not left the Corps to serve as *corregidor* of Oaxaca. Lafora's cheeky proposal met with the predictable denial. The viceroy had passed Lafora's petition on to the king, and his judgment on Lafora's final petition is the last piece of documentation on Lafora in the archives. The king informed Sabatini and Cavallero that army officers leaving their duties for political assignments were not permitted to return to their former military posts. In Lafora's case, his reinstatement in the Corps in his old rank and class would prejudice other members of the Corps, and therefore his petition was denied.[38]

Lafora was never readmitted to the Corps. The last time his name appeared on the general rolls was in 1774, when he was assigned to the *corregimiento* of Oaxaca. By 1775, he was no longer officially considered an engineer serving in New Spain: his name had been dropped from the official roster.[39] Lafora's days of service to the king as a royal engineer had ended when he accepted the civil post in Oaxaca. Although later he wished to return to his engineering

[38] BCM, vol. 57; AGN, *Indiferente de Guerra,* 331; Lafora to Cavallero, Mexico, Mar. 27, 1788; and royal order of Jan. 31, 1789 communicated in the king to Sabatini and Cavallero, the Palace, Feb. 7, 1789, AGM, *Expediente personal de D. Nicolás de Lafora.*

[39] "Lista general por antiguedad de los yndividuos del Cuerpo de Yngenieros, comprehendidos los de Yndias, con las variaciones ocurridas desde el mes de febrero anterior hasta Agosto de 1774," AGS, *Guerra Moderna,* 3794; and "Relación de existencia de los Ingenieros que sirven en el reino de Nueva España con expresion de sus Empleos, Destinos y Encargos," San Miguel de Perote, Mar. 24, 1775, AGN, *Historia,* 568.

vocation, stringent rules of the Corps prohibited him from doing so. However, leaving the Corps to be *corregidor* did not indicate official resignation, because throughout the fifteen years that Lafora served in Oaxaca and the subsequent years until his name disappears from the records, he was referred to as engineer in official correspondence.

Although Lafora had worked on engineering projects in Spain before he went to Mexico and although he worked in the capital area as an engineer, his fame and his best accomplishments are rooted in the Borderlands. His work was not forgotten. His diary of the Rubí inspection tour was reproduced many times, as mentioned above, and his general map of the Interior Provinces served as a model for later cartographic efforts. In 1778, another engineer, Luis Bertucat, drew a map based on the latitudes taken by Lafora a decade earlier.[40] A reproduction and simplification of Lafora's 1771 general map was made even later, in 1787.[41] Additionally, Lafora's map was to be reproduced in a universal history of North America that was planned and directed by four viceroys of New Spain during the closing years of the century.[42] Other fragments of Lafora's engineering projects in

---

[40] "Mapa del Derrotero que hizo el Cmdte. Gral, Cavo. de Croix por las provincias de su cargo desde la Ciudad de Durango hasta la Villa de Chihuahua, Año de 1778," AGI, *Audiencia de Mexico*, 538. Another copy of the map is in the SGE, 3-3-1-10; and it is published in *Cartografía de Ultramar*, vol. III, no. 11; and in Navarro García, *Don José de Gálvez*, following p. 176.

[41] "Mapa general de las provincias internas continentales," 1787(?), British Mus. Ms. Room, Add. 1765a; and reproduced in Navarro García, *Don José de Gálvez*, following p. 336.

[42] "Plan: Division y Prospecto general de las Memorias de Nueba España: que han de servir a la Historia Unibersal de esta Septentrional de America, 1791," MN, ms. 570 (Virreinato de Mexico IV).

the Borderlands include a listing for a plan of the province of New Mexico in 1779.[43]

But most of all, Lafora could not be forgotten because of his joint recommendations and tour with the Marqués de Rubí. Their efforts resulted in the establishment of the Commandancy General of the Interior Provinces. Along with other projects and reforms instituted by José de Gálvez, Lafora's and Rubís work shaped events in the Borderlands with an influence that has been felt to this day. While Lafora was with Rubí devising a new defense plan for the harsh, northern limits of New Spain, plans were begun and preparations were made for advancement of the frontier still farther north. Just as Lafora was to become known for the part he played in the tremendously important overhaul of defensive measures in the Borderlands, another engineer was to achieve even greater fame for his part in extension of the new California frontier.

---

[43] "Plano del Reino del Nuebo Mexico" in "Catálogo de los documentos . . . 1824," BCM, ms. 5-2-4-12. Along with the previously cited Lafora map of Lower California, this map has been lost or misplaced in the BCM or SGE.

4

Establishment and Defense
of Upper California

Lieutenant and assistant engineer Miguel Costansó was one of the first Europeans to lay eyes on the great estuary that is San Francisco Bay. As a member of a small detachment that journeyed overland from San Diego looking for the famed Monterey Bay, Costansó was one of the discoverers of that vitally important Borderlands defense post in 1769. The part Costansó played in the discovery and establishment of the California mission and presidio sites has made him the best known member of the Corps of Engineers in New Spain. But Costansó's short jaunt to Alta California in 1769-1770, was only the prologue to a long, successful, and eminently important career in the Corps of Engineers.

Born in Barcelona in 1741, Costansó entered the Corps when he was just 21 years old. His technical skills soon became apparent. In entrance examinations for the Corps he demonstrated superior talent in draftsmanship and fortifications theory, and was lauded by officials of the Corps.[1] During his first two years as second lieutenant, Costansó served in his native city, Barcelona, and then in Málaga, before being assigned to join the Villalba expedition to New Spain. Arriving in Veracruz in the spring of 1764, he was first assigned to the Gulf coast, under the immediate orders of Lieutenant Colonel Miguel del Corral. This man was Costansó's superior for many years to come, and heartily recommended Costansó for his services and promotions. Costansó remained in Veracruz until May of 1767, when, by his own petition, he was transferred to the command of Colonel Domingo Elizondo for the Sonora expedition. Details of Costansó's activities with Elizondo are undocumented, but his stay in Sonora was short. He was soon in Lower California doing reconnaissance map work,[2] and shortly after that, he was involved in plans for the Sacred Expedition.

Little doubt remains that the Visitor General, Don José de Gálvez, conceived the plan to settle Upper California. He had sent a special envoy to speak with Charles III and his chief minister, the Marqués de Grimaldi, about Gálvez' proposed settlements at Monterey harbor and other likely spots. Official approval for the venture arrived when Gálvez already had left Guadalajara for San Blas to prepare detailed plans for the California expedition.[3]

---

[1] Joseph Gandón to Marqués de Squilache, Málaga, Oct. 27, 1763, remitting Costansó's examination, AGS, *Guerra Moderna*, 3019.

[2] See above, p. 65.

[3] Priestley, *José de Gálvez*, pp. 246-48; see also Charles E. Chapman, *The Founding of Spanish California;* and Theodore E. Treutlein, *San Francisco Bay, Discovery and Colonization, 1769-1776* for the Gálvez plan to occupy California.

There, at the obscure and unlikely harbor of San Blas, Gálvez met in junta with prominent officials to lay specific plans for the expedition to Monterey. Among notables present were commandant of the navy and harbor of San Blas, Don Manuel Ribero Cordero; navigator and professor of mathematics, Don Antonio Faveau y Quesada; pilot Vicente Vila; and the engineer Don Miguel Costansó. Meeting on May 16, 1768, only three days after Gálvez' arrival in San Blas, and probably not much later than the engineer's arrival from the north, the junta drew up formal plans for the Sacred Expedition. Monterey would be occupied by land and sea parties in accordance with the royal order given by Charles III on January 23 of that year. Costansó was involved in the planning from the beginning.

The formal report of junta proceedings [4] includes mention made by the king in his *cédula* authorizing the Visitor General to take any steps he felt necessary in so important a project, and instructing him to send an engineer with the expedition to make exact observations and a map of Monterey Bay. This specific admonition by the king to include an engineer in the Sacred Expedition was demonstration of the crown's continuing interest in and respect for the Corps of Engineers. In having Costansó present at the junta, which must have been arranged before authorization and instruction from the king arrived, Gálvez had foreseen the importance of an engineer in preparing

---

[4] "Reporte de la Junta, en que se trató de la expedición de Monterey por Mar, y Tierra," San Blas, May 16, 1768, Huntington Lby., ms. GA 419; and translation, "Junta Held at San Blas" in Treutlein, *San Francisco Bay,* pp. 4-7.

California expedition plans. Costansó's contribution was great from the beginning; his participation was germane to the plan.[5]

The journal that Costansó kept on the voyage to California[6] has been considered authoritative since the day it was received in Mexico. This *Narrative* is a meticulous, trustworthy guide to the founding of the first missions and presidios in California. Its further value lies in the information Costansó himself gives on his own duties and activities as a corpsman.

Gálvez told the king that Alta California should be settled for defensive reasons. Imagined, or exaggerated though it may have been, the threat of Russian and British encroachment on New Spain's northern borders was seriously feared by the Spaniards. As part of the king's overall strengthening of military power and retrenchment in the new world, and to

[5] On the importance of Costansó at the junta, see Thurman, *San Blas,* p. 62.

[6] Adolph Van Hemert-Engert and Frederick J. Teggart, eds., "The Narrative of the Portolá Expedition of 1769-1770 by Miguel Costansó," *Publications of the Acad. of Pac. Coast Hist.,* vol. I, no. 4 (1910). This diary was first published in Mexico, dated Oct. 24, 1770, with the title, *Diario Historico de los viages de mar, y tierra hechos al norte de California,* MN, ms. 334 (Reino de Mexico II), also extant in other archives. There are many other editions of Costansó's *Narrative,* including Ray Brandes, tr. and ed., *The Costansó Narrative of the Portolá Expedition,* which contains a facsimile reproduction of the original printed copy, and bibliographical notes of other editions of Costansó's *Narrative,* plans, and maps.

The diary that Costansó kept on the land trip from San Diego to Monterey (parts of which were included in his *Narrative*), dated in San Diego, Feb. 7, 1770, is entitled, *Diario del viage de tierra hecho al norte de la California.* Manuscript copies exist in several archives, among them, an official copy dated in Mexico, June 20, 1770, in MN, ms. 333 (Reino de Mexico I), and AGN, *Historia,* 396. Several translations have been published. The best is Frederick J. Teggart, "Diary of Miguel Costansó," *Publications of the Acad. of Pac. Coast Hist.,* vol. II, no. 4 (1911).

protect against foreign intrusions, the king was sympathetic to California settlement. Costansó also recognized this point when he mentioned in the first paragraph of his *Narrative* that foreign intrusion was the reason and motivating force leading to the speediest possible occupation of California.[7] The engineer's value judgment, or explanation, on this much debated point is certainly worth serious consideration by contemporary scholars.

Members of the junta in San Blas had decided that the expedition would begin as soon as two packetboats, *San Carlos* and *El Príncipe,* returned from Guaymas where they were delivering troops and supplies to the Elizondo expedition, from which Costansó had just returned. Mission Santa María in Baja California was designated the rendezvous and starting point for the land expedition. Gathered there were forty men from the California Company and thirty Indian volunteers. Gálvez divided that force into two parties, one to march under Don Fernando de Rivera y Moncada, captain of the presidio of Loreto, with 25 soldiers and some friendly Indians. This first group was to march as advance guard and escort for cattle being driven to Alta California. The remainder of the men and provisions were to march under direction of Don Gaspar de Portolá, governor of California and commander-in-chief of the entire operation. Portolá's party left mission San Fernando de Velicatá, the jumping off point in Baja California, on May 15, 1769, accompanied by the famous Fray Junípero Serra, spiritual founder and first father president of the Upper California missions.

[7] Brandes, *Costansó Narrative,* pp. [21] and 78.

Meanwhile, the packet *San Carlos* arrived at La Paz in mid-December, 1768, and was ready to sail north after two weeks spent in repairs and provisioning. On board when it sailed on January 9, 1769, were 25 Catalonian Volunteers under Lieutenant Pedro Fages, Don Pedro Prat, a surgeon, Father Fernando Parrón, and the engineer Miguel Costansó. *El Príncipe* was to be commanded by Antonio Faveau y Quesada, "who would also act as engineer to help Costansó."[8] Treacherous waters at the lower end of the Sea of Cortés accounted for failure of the *San Antonio,* the name usually used for *El Príncipe,* to rendezvous with the *San Carlos* in the Bay of San Bernabé at Cape San Lucas. The two little ships were left to make their way separately up the Baja California coast to the port of San Diego, which was known to the Spaniards as the result of earlier navigational reports.

Costansó knew the problems involved in navigation of the outer coast of the peninsula. Prevalent north and northwest winds throughout the year blow directly contrary to the course of the voyage. The engineer noted in his *Narrative* that it was necessary to lengthen the trip by tacking westerly away from the

---

[8] Captain Vicente Vila kept a log of this stage of the Sacred Expedition. See Frederick J. Teggart and Robert Seldon Rose, eds., "The Portolá Expedition of 1769-1770: Diary of Vicente Vila," *Publications of the Acad. of Pac. Coast Hist.,* vol. II, no. 1 (1911).

Faveau was not a member of the Corps of Engineers, but "Was versed in that profession," being a mathematics professor. Treutlein, "Junta Held at San Blas" in *San Francisco Bay,* p. 7; Navarro García, *Don José de Gálvez,* p. 163. Faveau's map of the Gulf of California, reproduced in Navarro García, *Don José de Gálvez,* following p. 192, is noticeably inferior to Costansó's cartographical works.

coast, and then north. Upon reaching the desired latitude, quartering winds could then carry the ship into port. So, with orders to this effect, the *San Carlos* did sail west first, but was driven 200 leagues (about 500 miles) from the coast, much farther than was necessary or desirable, and consequently, ran short of fresh water. After an arduous voyage, the *San Carlos* finally arrived in San Diego on April 29, 110 days out of La Paz. The *San Antonio* had enjoyed better fortune since leaving the tip of the peninsula, and made San Diego in 59 days. Men on board both ships were in poor condition upon arrival. Two on each ship had died, most crewmen suffered from scurvy, including Costansó, and all were greatly weakened and tired by the trip.[9]

Landing in San Diego, officers Don Pedro Fages, Jorge Estorce, and Miguel Costansó led a party of 25 of the ablest men in search of water up Mission Valley. Costansó described what they saw in terms mixed with elements of scientific observation and lyric beauty. Practical data worthy of a botanist, geologist, or ethnologist, blended with the contemporary ecologist's awareness of natural beauty, fills all of Costansó's descriptions. When Costansó wrote that they saw a river (the San Diego) about twenty yards wide, bordered by lush cottonwoods, emptying into a lagoon that at high tide could accommodate the launch to take on fresh water, he does more than simply describe the area. He supplies precise scientific information of a technical nature, and at the same time, paints a literary picture of Arcadian di-

---

[9] Teggart, "Narrative," pp. 24-27.

mensions. He noted the presence of wild grapevines, also mentioning that they were covered with tender blooms and very lovely. Sweet-smelling rosemary reminiscent of Spain, and other herbs were mentioned in the same breath with the natural fertility of the soil, once again combining the aesthetic with the practical.

Costansó also described natives and wildlife encountered in his meanderings. Weapons, clothing, tools, skills, and appearance, as well as disposition of the Indians, were chronicled with the thoroughness of a scientist. He said that the Indians made much of their bravery, although in fact they exhibited but little; that they were overbearing, tricky, covetous, and insolent. All of these were character traits that later Spaniards observed and were forced to deal with. Costansó described an abundance and variety of fresh and salt water fish and shellfish, and much wild game and birdlife, again showing a keen awareness of his surroundings and an appreciation of future needs of Spanish colonists.[10]

While Costansó had been busy observing and describing the area of the port of San Diego and recording data for his reports,[11] the land parties from Santa María in Baja California arrived. Though ill

[10] Teggart, "Narrative," pp. 27-33.

[11] Costansó prepared a map of the area reconnoitered and described so eloquently by him, but it has been lost. Brandes, *Costansó Narrative,* p. 12; and Costansó to Gálvez, San Diego, June 28, 1769, in the Huntington Lby., cited in Brandes, *Costansó Narrative,* p. 18, n. 6. Notes made by the engineer on this reconnaissance are listed as "Relacion del Viaje de la cañada de San Diego a San Diego," in Vivian Fisher, "Key to the Research Materials of Herbert Eugene Bolton," Bancroft Lby., C-B840. Hereinafter cited as BL.

with scurvy or some other dietary deficiency, the engineer accompanied Lieutenant Fages, Don Gaspar de Portolá, and six soldiers north in search of the main object of the expedition: Monterey Bay. In the passage north, closely following the seashore, Costansó described the actions, language, clothing, crafts, and food of the Chumash Indians, along the Santa Barbara Channel. He was extremely complimentary in this description, noting a higher degree of civilization among the Channel Indians than those near San Diego. As examples of the more complex culture, he cited the existence of both eunuchs and homosexuals, and a considerable division of labor.

At this point, great confusion between previous descriptions and contemporary latitude readings for the bays of Monterey and San Francisco come into the *Narrative*. Brief and ambiguous landmark notations made by Joseph González Cabrera Bueno in his navigation handbook of 1734,[12] matched what Costansó saw with his own eyes at Monterey Bay, but the latitude was wrong. After making their way farther north, members of the party realized that what is now known as Drake's Bay, was considered by Cabrera Bueno to be San Francisco Bay. Cabrera Bueno had missed completely the great inland estuary behind the fog-shrouded Golden Gate.

What Costansó and his fellow journalist Father Juan Crespi saw was conditioned by what they had read in Cabrera Bueno's guide. When they reached the ridge on the present San Francisco peninsula behind Half Moon Bay, and were able to spot the

---

12 Joseph González Cabrera Bueno, *Navegación Especulativa y Práctica*.

Farallones and Punta de los Reyes; they knew that Cabrera Bueno's landmarks for the location of San Francisco Bay were correct, and that Monterey lay behind them. To their right was something new. The estuaries of present San Francisco Bay were explored hastily by a small detachment. Costansó noted the immensity of the bay, but did not speculate on its possibilities for settlement or as a defensive bastion, undoubtedly because orders specified the founding of a mission and presidio at Monterey, which the party then knew lay to the south. Orders were to be obeyed first, and questions asked later.[13]

With great hardship caused by food shortage, the little party that had gone north overland and first sighted San Francisco Bay, returned to San Diego after having waited ten days at Monterey for a ship to arrive. In San Diego, matters were desperate. Supplies were low, men were sick, and no ship had arrived from San Blas carrying provisions, news, or encouragement. Portolá sent a detachment of forty men to retrieve the supplies and cattle that had been left at the jumping off point of Velicatá. The next month, in March of 1770, the *San Antonio* arrived in San Diego, as if miraculously in answer to Father Serra's *novena*. Bolstered by provisions and renewed hope, Portolá sent twenty men under Fages from San Diego to Monterey by land. At the same time, on April 16, others, including Father Serra and Cos- tansó, boarded the *San Antonio* and sailed north for Monterey. Both parties arrived safely by the end of May. This sojourn gave Costansó an opportunity to

---

[13] Teggart, "Narrative," pp. 37-55; Treutlein, *San Francisco Bay*, p. 16.

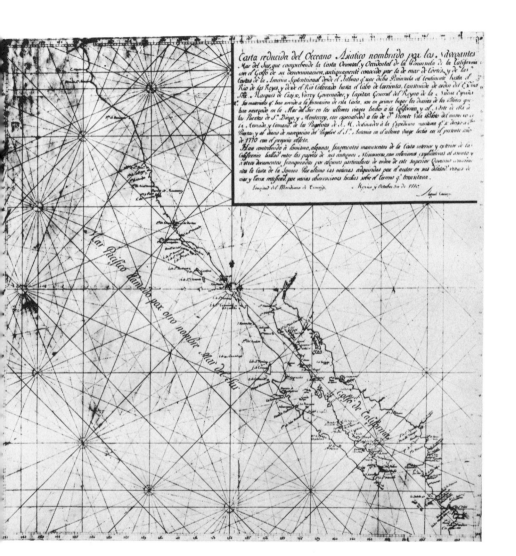

MAP OF THE PACIFIC COAST
Miguel Costansó's manuscript copy drawn in 1770.
See page 107. Courtesy, Bancroft Library.

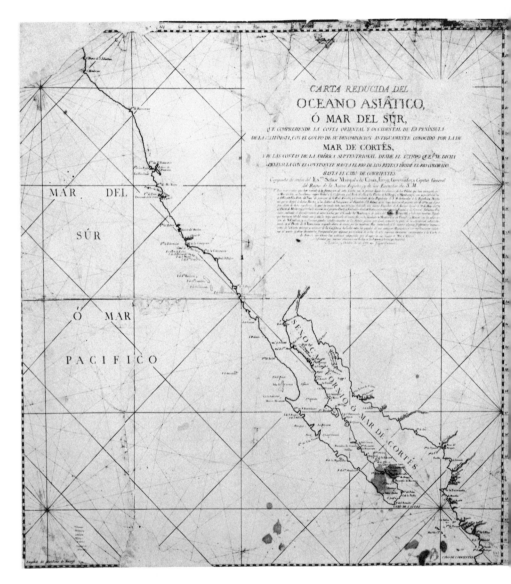

MAP OF THE PACIFIC COAST
The printed edition of Costansó's 1770 map.
See pages 107 and 103. Courtesy, Bancroft Library.

take careful note of Monterey and its surroundings and to insert a lengthy description in his *Narrative*.

Costansó depicted surrounding slopes, opportunities for anchorage, depths of the bay floor, shelter and direction of the coast, and other important details. In his usual manner, Costansó did not neglect aesthetic considerations. He described pines and other trees on the ridge and foothills of the Sierra de Santa Lucía, as well as springs, rivers, lagoons, plentiful wildlife, and gentle Indians. The engineer spent a little over three months in Monterey, and during that time he kept busy. He chose the site for the presidio of San Carlos de Monterey, as well as for the mission, its out-buildings, and offices. The engineer utilized his architectural skills in tracing out lines for the foundations of the buildings and defense works.[14]

Before Costansó left Monterey on July 9, 1770, aboard *El Príncipe,* he had seen to the construction of two storehouses where cargo from the ship was stowed, and which served as provisional living quarters for the missionaries and the commanding officer. A third storehouse was built to hold powder and defense equipment, at a safe, but visible distance from the other supplies and habitations.[15]

Besides supervising the first European construction at Monterey, Costansó was busy drawing plans and

[14] Costansó to Marqués de Croix, San Blas, Aug. 2, 1770, AGI, *Audiencia de Guadalajara,* 417. The unsigned plan entitled "Plano del Real Presidio de Sn. Carlos de Monterrey," reproduced in Irving B. Richman, *California Under Spain and Mexico, 1535-1847,* opposite p. 338, appears to be Costansó's work.

[15] Costansó to Gálvez, San Blas, Aug. 2, 1770, AGI, *Audiencia de Guadalajara,* 417.

maps, and taking notes for later reports. He made a sketch of the port and land surrounding it, which was still in first draft when he arrived in San Blas. Costansó explained to Viceroy Croix and to Gálvez when he wrote to them from San Blas upon arrival there on August 2, that he had had neither time nor place to complete the maps he had begun. He worked most of the day outside, supervising construction, and at night did not have sufficient light or clean space and quietude to complete the close work. But he promised both officials personal delivery of the documents he had prepared when he met them in Mexico City.[16]

Costansó had spent fourteen months and ten days in Alta California, and in that time had been a witness and a contributor to the first European colonization activities in the area. The engineer never returned to the pine covered slopes and the wide spacious harbors of California that he described so well; he was never again to meet the docile California Indians whose habits he chronicled so completely. But the reports he wrote, the maps and plans he drew, and his unquestioned understanding of the area were to qualify him as an expert on California throughout his long career in Mexico.

Besides Costansó's *Narrative* and *Diary* and the letters he wrote describing his activities in California, several maps have survived as evidence of his work in the new province. The most notable map was made in two very similar versions, both dated

---

[16] Costansó to Croix, and Costansó to Gálvez, San Blas, Aug. 2, 1770, AGI, *Audiencia de Guadalajara,* 417.

October 30, 1770,[17] no doubt designed to accompany the *Narrative* dated less than a week earlier. Both versions have long titles differing only slightly and show the coast of California from the forty-third parallel, north of Cape Mendocino, to the twentieth parallel, including the port of La Paz and Cabo de Corrientes. Both coasts of the Gulf of California are delineated with more than 130 important place names marked on all coasts. Costansó listed the various sources he used in compiling the map, such as diaries of various pilots, especially Vicente Vila, commander of the *San Carlos* on the Sacred Expedition; papers from missionaries in Lower California; government documents; and data collected by Costansó himself. The map has rhumb lines, as well as latitude and longitude measured from the Meridian of Tenerife, Canary Islands, the base meridian most commonly used by Spaniards during this period. The map is aesthetically pleasing and quite accurate. Both versions were printed in Madrid in 1771.

---

[17] "Carta reducida del Oceano Asiatico nombrado por los Navegantes Mar del Sur que comprehende la Costa Oriental, y Occidental de la Peninsula de California con el Golfo de su denominacion, antiguamente conocido por la de mar de Cortes, y de las Costas de la America Septentrional desde el Isthmo que une dicha Penensula al Continente hasta el Rio de los Reyes, y desde el Rio Colorado hasta el Cape de Corrientes, construida de orden del Exmo Sor Marqués de Croix . . . Mexico, Octubre 30, de 1770, Miguel Costansó," SGE, J-9-2-e, reproduced in *Cartografía de Ultramar,* vol. II, no. 127; and "Carta Reducida del Oceano Asiático, o Mar del Sur, que comprehende la costa oriental y occidental de la peninsula de la California, con el golfo de su denominacion antiguamente conocido por la de Mar de Cortés, y de las costas de la America Septentrional desde el Istmo que une dicha Peninsula con el continente hasta el Rio de los Reyes, desde el Rio Colorado hasta el Cabo de Corrientes. Compuesta de orden del Exmo. Senor Marques de Croix . . . , Mexico, y Octubre 30 de 1770, Miguel Costanso," BCM, no. 4934. *See* pp. 103-04.

Costansó also drafted other less important and less impressive maps while he was in Alta California. One showed the area from Cape Mendocino at forty degrees north latitude to San Blas at twenty-one degrees.[18] It includes the Bay of La Paz and Cerralvo on the Gulf side of Lower California but, other than those points, does not delineate the interior coast. Another map of the Port of Monterey [19] in watercolor includes soundings of the bay, particular areas labeled, and a good representation of the surrounding terrain which complements the *Narrative* description of Monterey.

Shortly after arrival in San Diego in 1769, Costansó had been promoted to lieutenant and extraordinary,[20] and back in Mexico after the Sacred Expedition, he was raised in army rank to captain in recognition of his services.[21] In the capital, Costansó began work on innumerable projects of civil engineering and architecture for that area that were to occupy most of the remainder of his career. But his departure from California in 1770, and subsequent occupation with other tasks by no means signaled

18 "Plano de la Costa del Sur correxido hasta la Canal de Santa Barbara en el Año de 1769," SGE, J-9-2-c, and reproduced in *Cartografía de Ultramar*, vol. II, no. 126.

19 "Plano del Fondeadero, ó Surgidero de la Bahia, y Puerto de Monterrey, situado por 36 grados, y 40 minutos de Latitud Norte, y por 249 grados, 36 minutos de Longitud, contados desde el Meridiano de Tenerife," original in SGE, LM-8-1-a, and reproduced in *Cartografía de Ultramar*, vol. II, no. 128. Several other Costansó maps of the Lower California coast and other places related to the new California settlements are listed in Brandes, *Costansó Narrative*, pp. 111-12. *See also* frontispiece, herein.

20 Service record, Mar. 24, 1775, AGN, *Historia*, 568.

21 Juan Gregorio Muniain to Julián de Arriaga, the Palace, Jan. 5, 1771, remitting royal orders for promotion to captain of Costansó and Pedro Fages, AGI, *Audiencia de Guadalajara*, 417.

the end of Costansó's interest or participation in development of the new province. All the while Costansó lived at the viceregal court, from 1770 until his death in 1814, he was consulted as an expert on the northern or California frontier and was called upon for his educated opinions on defense questions and the problems attendant to defensive expansion.

Costansó's next commission concerning the Borderlands, while he was not actually in the area, was in 1772. Viceroy Antonio María Bucareli y Ursúa called Costansó to a junta which met on October 17. The object of the meeting was to discuss the feasibility of a land route between Sonora and Monterey that had been petitioned by Juan Bautista de Anza, captain of the presidio at Tubac. Costansó and other members of the board of experts concurred that Anza's proposal was realistic.[22] The engineer had no doubt that exploration would be beneficial to the struggling California settlements. Supply from Lower California was uncertain: by land, the trip was long and hard, and by sea, particularly dangerous in certain seasons. Additionally, the barren Baja California peninsula furnished little aid itself. It was only Sonora that could save Alta California.[23]

Costansó's favorable confirmation of the suitability of the land route helped assure Anza's fame. The presidio captain was given clearance to lead his now famous overland expeditions in 1774 and 1775. Based on advice from experts like engineer Costansó, Buca-

---

[22] Bucareli to Arriaga, Mexico, Oct. 27, 1772, AGI, *Audiencia de Guadalajara*, 513.

[23] Herbert Eugene Bolton, *Anza's California Expeditions*, vol. I, pp. 45-50.

reli supported the ventures which resulted in Anza's colonizing treks and the founding of San Francisco in 1776.[24] And during the time Anza was marching from Tubac to San Francisco, Costansó was busy with yet another matter concerning communication between the older Borderland areas and California. By order of the viceroy, the engineer prepared a report in the spring of 1776 dealing with the distance between Santa Fé and Monterey, and from the New Mexican capital to Tubac.[25]

Costansó characteristically credited very carefully the sources upon which he based his estimates. As in other work he produced, the engineer relied upon data from others as well as figures that he himself had collected. In his 1776 report, Costansó used material supplied in letters from Pedro Fermín de Mendinueta, governor of New Mexico, and Father Silvestre Vélez de Escalante, missionary at Zuni. He also utilized his own astronomical observations and those of others mentioned in the report. In a very orderly and scientific manner, Costansó gave the latitude and longitude of Mexico City and Cape San Lucas, at the southern tip of Baja California. He then followed with the longitude and latitude of Monterey and from this knowledge computed the direct distance between Monterey and Mexico as 705 leagues.

[24] Chapman, *A History of California: the Spanish Period,* pp. 297-98.

[25] "Informe de Don Miguel Costansó sobre la distancia que media de la Villa de Santa Fe del Nuevo Mexico y la Sonora y entre aquella Villa y Monterrey, 1776," Biblioteca Nacional (Madrid), ms. 19266. Hereinafter cited as BN. This document has been published in Spanish in *Noticias y Documentos Acerca de las Californias, 1764-1795* of the Colección Chimalistac, vol. v; and there is a manuscript copy entitled, "Informe que D. Miguel Costanzó rindio al Virrey Bucareli, acerca de las distancias que hay entre Nuevo Mexico, Sonora, y la Nueva California, Marzo 18, 1776," AGN, *Provincias Internas,* 169. See Appendix C.

From this point, Costansó had to rely on conjecture and estimate, explaining that exact measurement was unknown and apologizing to the viceroy for imprecise but necessary guesswork. By using known latitudes and longitudes, he was able to calculate exact distances between the various spots. Then, by adding on a reasonable distance for fording rivers and other natural obstacles encountered in cutting a new route, the engineer came up with estimates of 300 leagues from Santa Fé to Tubac and 500 leagues from Tubac to Monterey.[26] Before ending his report, Costansó warned that the distances he computed were the very shortest possible and that anyone undertaking such a trip should be forewarned that the journey might be longer than expected.

Connecting Spain's two most northerly outposts, New Mexico and California, was a major defense and economic project that Bucareli encouraged. While Anza continued westward to California, Father Francisco Garcés explored north up the Colorado River and to the Hopi villages toward Santa Fé. Concurrently, Father Escalante and Father Francisco Atanasio Domínguez were tredding a path later known, in part, as the Spanish Trail, in one more attempt to link Santa Fé with Monterey. Costansó, in his 1776 report, was therefore intimately involved in Bucareli's ambitious plan to establish land connections between the two northern enclaves of Spanish dominion, the newest western Borderlands capital at Monterey, and the oldest at Santa Fé.

---

26 Shortest major highway access routes in 1975, show Santa Fe to Tubac 704 miles, and Santa Fe to Monterey 1,118 miles. Using 2.3 miles to the league, Costansó estimated 690 and 1,150, respectively, for the two routes, showing a commendable accuracy in his calculations.

Shortly after the 1776 report, Viceroy Bucareli again consulted Costansó about another California problem. At this time, Bucareli was concerned about the silting up of the harbor at San Blas, which jeopardized shipping to the Alta California settlements. Despite his participation in two separate meetings on the subject, the engineer's advice on this matter was not taken. Both Costansó, who had mapped the harbor and knew it well,[27] and Ignacio Arteaga, ranking naval officer at San Blas, thought that the naval depot should be transferred from San Blas to some other harbor, perhaps Acapulco, because of the inadequacy of the former port. But Costansó's commanding officer in the Corps, Miguel del Corral, and other officials disagreed, and San Blas remained the principal but still inadequate port on Mexico's west coast north of Acapulco.[28]

Costansó had reason to know something about harbor defense and fortification construction. He had served in Barcelona and Málaga in Spain and in New Spain at Veracruz. Additionally, Viceroy Bucareli sent Costansó to Acapulco, the principal Pacific port of entry, in 1776, to reconnoiter and plan reconstruction of the Castillo de San Diego at the port which had been damaged severely by an earthquake on April 21, 1776. Out of this reconnaissance, Costansó prepared a lengthy report and plans for the

---

[27] Costansó's map of the naval depot harbor was made when he was at the port for the junta planning the Sacred Expedition. "Plano del Puerto y Nueva Población de San Blas Sobre la Costa de la Mar del Sur," San Blas, May 23, 1768, SGE, J-3-2-48; and reproduced in Thurman, *San Blas,* p. 63.

[28] Thurman, *San Blas,* pp. 223-40; Bernard E. Bobb, *The Viceregency of Antonio María Bucareli,* pp. 169-71.

amplified harbor fortification and defense works.[29]

Costansó's next important involvement in Border-lands defense came many years later. Predictably, once again, Costansó was consulted on California; his presence at the founding of the new province and his many years of consistently outstanding service qualified him as one of the few trusted technicians for protecting California. By the 1790s, Spanish officials in Mexico and Spain were suffering from intensified paranoiac fear that California, their buffer Border-land province, was to be wrested away from them by the imperialistic Russians or English. Certain activities of Costansó were just as much a part of this colonial international drama as the well known show-down between England and Spain at Nootka Sound.

In November of 1792, the viceroy, the second Conde de Revilla Gigedo, wrote a long letter to the king in which he listed the many ills of California and his suggestions for remedies.[30] Foremost, he noted that the presidios at San Diego, Monterey, and San Francisco were totally incapable of repelling an attack from foreign forces. The viceroy compli-mented the troops as being more than adequate to

---

29 "Expediente sobre el proyecto de las obras de defensa en el Castillo de San Diego," signed by Miguel Costansó in Acapulco, May 22, 1776, BCM, ms. 5-3-10-7. Unsigned plans in Costansó's handwriting, which were to accompany the report, have also been located: "Plano del Terreno que ocupa el Castillo de Sn. Diego del Puerto de Acapulco del Proyecto formado para maior seguridad y defensa de este importante Puerto," "Proyecto de un Castillo para defensa de la entrada al Puerto, y Poblacion de Acapulco" (draft and final version of front elevation and profile views), and "Pro-iecto de un Castillo para defensa de la entrada del Puerto de Acapulco" (draft and final version of plan view), SGE, K-b-5-8.

30 Conde de Revilla Gigedo to Conde de Aranda, Mexico, Nov. 30, 1792, AHN, *Estado,* 4290.

deal with hostile Indians but pointed out that they had neither training nor equipment to fight Europeans. Artillery was lacking that could in any way defend the harbor entrances; and though the viceroy realized that the California presidios need not be impregnable fortifications, they required additional batteries for even minimal defense capability. Revilla Gigedo also thought that California needed a more experienced man as governor. José Joaquín Arrillaga was then serving an interim appointment. One of Revilla Gigedo's more notable suggestions was that he send one of the best engineers in the New Spain contingent to California to supervise establishment of improved fortifications. The viceroy knew that financial subsidy from the crown would be necessary to cover the costs of sending an engineer, and he closed his letter imploring the king to give special attention to this idea.

Very likely, Revilla Gigedo was considering sending experienced Miguel Costansó to Alta California. After all, he was already familiar with the area and its problems, and was a high ranking member of the Corps of Engineers in New Spain by 1792.[31] But Costansó was not sent to California for a second time, nor was any other military engineer just yet. Revilla Gigedo received a reply assuring him only that the king was considering his important recommendations with the care they deserved.[32]

Revilla Gigedo was not to see the result of his

---

[31] In 1790, Costansó was fourth in rank in New Spain; by 1795, he was second. AGS, *Guerra Moderna,* 5837.

[32] [The king] to Revilla Gigedo, Aranjuéz, Feb. 23, 1793, AHN, *Estado,* 4290.

request for aid to California. His successor, the Marqués de Branciforte, continued the effort to put California's defenses in proper order, possibly jolted into action by George Vancouver's visit to San Francisco bay in 1792 and 1793. By decree of September 20, 1794, the viceroy ordered Costansó to state his opinions on the strengthening of the California presidios.[33] Costansó's prompt report, dated October 17, 1794, is an extremely important document. Not only did the engineer point out the defense problems, which had already been recognized by Viceroy Revilla Gigedo from reports he had received but the engineer also made two other major points besides actual recommendations for improving the defense situation. Costansó recognized the British as a vital and expanding colonial power of genius, and he recommended populating California with Hispanicized people from New Spain.

Costansó began his report to the viceroy by explaining the existing state of defense in California. There were four presidios garrisoned by 218 men expected to defend 520 miles of coastline from San Diego to San Francisco. These men were supposed to deal with reduction and control of the natives, and also to maintain good will between mission establishments and hostile Indians. Adequate defense against foreign invasion was impossible. An expenditure of funds

---

[33] Marqués de Branciforte to Duque de Alcudia, Mexico, July 3, 1795, AHN, *Estado,* 4290; and "Informe de D. Miguel Costansó al Virrey el Marqués de Branciforte, sobre el proyecto de fortificar los presidios de la Nueva California," BN, ms. 7266; also available in BL, *Mexican Mss.,* 401; published in Spanish in the Colección Chimalistac, vol. v; and translated by Manuel P. Servín, "Costansó's 1794 Report on Strengthening New California's Presidios," *Calif. Hist. Soc. Qtly.,* vol. XLIV, Sept. 1970, pp. 221-32.

had to be made to make California defensible. For direct military resistance, Costansó recommended batteries of eight twelve-pound cannon with parapets built of dirt covered by adobe bricks costing, he calculated, 8,000 pesos each. New gun emplacements were useless without trained artillerymen, so Costansó further recommended importing and stationing trained gunners at the new batteries.

But Costansó realized that added batteries to defend against naval attack was only part of a greater solution. What really had to be done was to find a means of dealing with foreigners infringing on Spanish California lands. He summarized the activities of English fur trappers and traders north of the settlements. Costansó observed that despite what the Duque de Almodóvar, minister of the king to the court of St. Petersburg, had said (in a letter quoted by Costansó) about the remoteness of a Russian invasion on Spanish territory, there was still a great deal to fear and from a much more awesome power than Russia. Great Britain, Costansó objectively and astutely perceived, had a special facility for colonization and commerce. This power was sure to extend its then still small Pacific coast operations into major colonies, possibly rivaling Spanish Manila trade by using Macao or Canton as commercial outlets.

To meet the English challenge, Costansó suggested strongly to the viceroy that continuance of a policy of indifference or inaction would result in disaster for the glorious name of Spain. Even with the batteries that he suggested, Costansó said that coastal defense would not be entirely adequate, because California lacked a sufficient number of Spanish colonists to lend auxiliary support and supplies in case of attack. The first thing to do then, was to populate the country. Costansó pointed out that the missions and missionaries of California, while serving His Majesty honorably in converting the natives to the Catholic faith, continued to draw on the royal treasury. The best way to accomplish two goals at the same time, according to Costansó, was to introduce into California settlement *gente de razón,* or people of Spanish culture, either peninsular Spaniards or Creoles, or mestizos accustomed to the Spanish way of life. In this way, the Californias would become more populated and therefore more capable of defense; at the same time, skills, arts, and crafts would be taught to the Indians by example, and the establishments also would become economically independent.

A further suggestion made by Costansó was that the annual trading ship sailing from Cavite in the Philippine Islands should make a stop at the port of Monterey. This would serve to refresh the sailors aboard and at the same time provide trading opportunities for luxury-starved residents of Upper California. Some of the sailors, seeing the beauty and attractiveness of California, might want to stay there either as soldiers or colonists.

Above all, Costansó said, it was important to promote navigation and commerce in the waters along the coasts of Sonora, Nueva Galicia, and the Californias. He advocated giving residents permission to build small boats and the privilege of free trade among themselves. This would serve to promote the general prosperity of those provinces, which chronically suffered from high prices and shortage of goods. Trade items still had to be shipped all the way from the capital or even from Spain. There were many articles, such as fish, available only in Upper California waters, that would be most desirable to residents of lower coastal areas. Restricted coastal trade in California was indeed instituted in later years, and increasing American smuggling proved that enlarged commercial activities were economically stimulating, as Costansó had predicted.[34]

Costansó closed his report by saying that his suggestions, if implemented, would be sufficient to make California bloom and flower into a prosperous and defensible colony. Nothing quite so rosy resulted in California, but then not all of Costansó's recommendations were put into operation. At least one of his remedies was applied very quickly however, as reported to the king by Viceroy Branciforte in July of 1795.[35] The viceroy said that besides the advice proffered by Costansó, artillery commandant Brigadier Pablo Sánchez had suggested the construction of entrenchments or strong points in each presidio, and the formation of a fixed provincial artillery company for California, with its base and greatest complement located at Monterey. Governor Diego Borica in California had reported the completion of the entrenchments at San Francisco and was working on a battery of eight cannon at the entrance to Monterey at Costansó's suggestion. Awaiting further

---

[34] Thurman, *San Blas,* pp. 358-59.
[35] Branciforte to Alcudia, Mexico, July 3, 1795, AHN, *Estado,* 4290.

orders from the king, Branciforte would continue to study the problem.

Later the same month, the viceroy wrote to the king that he had convened a special commission of consulting chiefs to formulate a *dictamen* on what specifically ought to be done to defend California. On July 13, 1795, three experts signed a document they had prepared for the viceroy. Predictably, Costansó, by then lieutenant colonel and second class engineer,[36] was one of the highly respected threesome. The other members were Pablo Sánchez, the artillery commandant consulted previously by Branciforte, and Salvador Fidalgo, frigate captain of the royal navy just returned from Nootka and California. Their report[37] is a very specific and precise instruction to the viceroy of what exactly ought to be done to defend California considering the funds available.

First of all, Sánchez, Fidalgo, and Costansó warned that sending aid to California in the form of population, artillery, and military supplies was in itself a difficult and arduous undertaking. They pointed out that the principal difficulty was the shortage of ships to carry the passengers and supplies, and recommended sending two or more transports instead of risking the dangers of overcrowding and overloading one vessel. As for specific points of defense, the trio

---

[36] Service record, 1795, AGS, *Guerra Moderna,* 3794.

[37] "Informe de Pablo Sanchez, Salvador Fidalgo y Miguel Costanso, sobre el proyecto de enviar auxilios a la California Alta. Mexico, 13, de julio de 1795," certified copy dated July 31, 1795, AHN, *Estado,* 4290; and "Consulta de Dn Pablo Sanchez, Dn Salvador Fidalgo, y Dn Miguel Costanso; hecha al Exmo Sor Virrey Marques de Branciforte sobre los auxilios que se proponia embiar a la California," Mexico, July 13, 1795, BL, *Mexican Mss.,* 401. See Appendix D.

said that only the three principal ports should be considered because of financial limitations. Construction already begun on the San Francisco battery was wisely ordered because of the size, location, and natural advantages of that port. The battery still needed men and ammunition to service it, since the twelve cannon were there but without shot and powder. No defenses whatsoever existed at San Diego and Monterey. Costansó, Fidalgo, and Sánchez in their *dictamen,* urged the construction of batteries in those harbors, all minutely specified.

Very carefully Sánchez, Fidalgo, and Costansó explained that the defense projects they proposed would only provide protection from corsair attacks; a major assault including landing forces could not be repelled. In the event of an invasion in force, it was suggested that the commanders retire their forces with the population and livestock to the interior, and from there harass their attackers as best they could using guerrilla tactics. As it happened, California was never invaded by a large military force; nevertheless, the report had dealt with this possibility and offered a viable solution. Indeed, this became the permanent plan for threatened attack and was carried out under Governor Pablo Vicente de Solá in 1818 at the time of the Hippolyte Bouchard corsair attack on the coast.[38]

The experts' report explained in detail their prescription for California defense. Eight men were required to service each artillery piece, and so a total

---

[38] Chapman, *Spanish California,* pp. 443-49; and Richman, *California,* pp. 211-13.

of 160 artillerymen would be needed for the twenty cannon distributed among the three presidios. But by adapting defense to the expected corsair type attack, forces could be cut in half, requiring thirty-two men each at San Francisco and San Diego, and sixteen at Monterey. The experts thought that eight or ten trained artillerymen among the eighty would provide a cadre to each proper handling and procedures to others. It was recommended that all the men be married, that they take their families with them, and that they settle there permanently. Supplying the needed artillery crews while augmenting California's population bears the distinctive Costansó ideology for aid to California. Further, the viceroy's three advisers suggested high standards for the soldiers chosen to man the batteries. It was recommended that they be subject to the same duties as other soldiers while not operating or caring for the guns and that they receive regular pay of their rank and assignment.

After proffering weak apologies for not being able to come up with any more impressive suggestions because of financial limitations, the three strongly suggested that at all times two ships should be at San Blas assigned exclusively to reconnaissance and transportation along the coasts of California. Only in this way, they said, could the viceroy keep abreast of what was going on in the province.

Two weeks after Branciforte's advisers had submitted their report to him, the viceroy had already taken affirmative action on their recommendations.[39] He ordered Lieutenant Colonel Pedro Alberni to

---

[39] Branciforte to Alcudia, Mexico, July 31, 1795, AHN, *Estado,* 4290.

pick seventy-two of his best married men from his First Company of Catalonian Volunteers, then at Perote, and dispatch them to San Blas for transportation to California.[40] The viceroy also ordered Pablo Sánchez to select one sergeant, three corporals, and fourteen soldiers from his artillery companies to be sent to California. Besides arranging for these additional troops, Branciforte ordered from San Blas and Acapulco artillery pieces, shot, powder, and other equipment needed to provide defense works as recommended by the *dictamen*. The viceroy prepared all the required orders for transfers, ships, and supplies to take the supporting force to California. He gave orders in preparation of inaugurating frequent naval reconnaissance of the California coasts. He instructed the governor of California on how to deal with foreigners arriving in Spanish ports and told him that if troops or supplies were needed in an emergency, he should request them directly from Pedro de Nava, Commandant General of the Interior Provinces.

In reporting all this to the king, Branciforte explained that the royal treasury would be depleted very little by all the provisions he had made. Money

---

[40] Although Branciforte's letter, cited above, does not mention ordering Alberni to California, he did so. Alberni was sent to California as military commandant of San Francisco. He took with him 78 of his Catalonian Volunteers, not merely 72 specified by Branciforte according to his own letter to the king. See Thurman, *San Blas,* p. 355, n. 11. In another letter, the viceroy said he had sent the company of Catalonian Volunteers and a detachment of eighteen artillerymen. Branciforte to Prince of Peace, Mexico, Nov. 30, 1795, AHN, *Estado,* 4290. Richman states that Alberni took 75 of his men to California, and that they arrived there in 1796 or 1797, along with the eighteen artillerymen under Sergeant José Roca. See Richman, *California,* p. 170. These figures agree with material in BL, *Calif. Arch.,* 8.

would be expended only for transportation costs and the slightly increased salaries awarded soldiers for duty in distant and expensive California. Through all that he had put into effect, California could be defended adequately against small scale and corsair attacks, and would be in the best possible state of defense excluding construction of huge fortifications and other financially impossible measures.

The extra soldiers that Costansó and his associates recommended were indeed sent to California, and with them went extraordinary engineer, Captain Alberto de Córdoba, just returned from duty in the Philippines. While making other arrangements for the defense supply of California, Viceroy Branciforte wrote to Governor Diego Borica explaining his belief that the services of an engineer were essential for organizing California's defenses. Accordingly, the viceroy saw to it that the director of the Corps appointed an engineer for California, the first since Costansó had been with the founding party in 1769 and 1770. Branciforte informed Borica that Córdoba was named for the California post, and that the governor should assign various projects to Córdoba wherever he was needed.[41]

Córdoba arrived at the presidio of San Diego on December 11, 1795.[42] In California for almost three years, the engineer was kept extremely busy by orders of the governor. Córdoba's activities can be divided into three general areas. Throughout his stay, he

41 Branciforte to Borica, Mexico, Oct. 17 and July 26, 1795, BL, *Calif. Arch.,* 7.

42 Antonio Grajera to Borica, San Diego, Dec. 17, 1795, BL, *Calif. Arch.,* 55.

supervised strengthening of the San Joaquín battery at the Golden Gate and engaged in other similar defense improvement projects at other harbors; and he wrote reports and prepared maps showing the state of California settlement after the first quarter-century of Spanish occupation. Additionally, Córdoba supervised the founding and establishment of a new town, the ill-fated Villa de Branciforte.

Córdoba spent some months in the south before presenting himself to the governor in Monterey or beginning his work on the San Francisco harbor fortifications. Once arrived at the most northerly settlement, Córdoba's advice was anxiously solicited for the repair of San Joaquín after a storm had damaged the little fortress in February of 1796. Also urgently needed was Córdoba's aid in repairing the powder storehouse at San Francisco, where a breach in the adobe wall threatened "catastrophe" from the strong winds that blew sand into the structure. The tower and sentry-box of the presidio also required Córdoba's attention during his first California spring. The roof of the presidio was also in need of repair.[43]

These relatively small projects in San Francisco and probably similar ones in Monterey on his way north, familiarized Córdoba with the all too depressing facts about the state of California's defense. In September of 1796, after less than a year in California, Córdoba wrote a report on the fortifications defending Spain's northern Pacific coast settlements.[44] His report shows that maintaining the harbor fortifi-

---

[43] Borica to Córdoba, Monterey, Mar. 6, 1796, BL, *Calif. Arch.*, 14.

[44] "Informe sobre fuertes," Monterey, Sept. [n.d.] 1796, BL, *Calif. Arch.*, 8. See Appendix E.

cations as they were was all but completely useless, because they were incapable of repelling even a small corsair attack. At the Battery of San Joaquín, guarding San Francisco, the structure itself was incredibly weak. Frequent earth tremors shook the building and the engineer feared that a small earthquake would force sea water to undermine the foundations. Likewise, even the cannon salutes fired at ships entering the harbor made the walls of San Joaquín shudder and shake. Besides needing structural strengthening, San Joaquín lacked effective firepower. There were only two cannon facing toward the west, or the approach to the harbor, and their projectiles did not cross. Making the situation worse was the fact that artillery range of any of the thirteen mounted pieces did not span the mouth of the bay. An enemy ship could safely enter within sight of the battery that was supposed to defend the opening. The artillery was too small, the engineer said, and in bad condition. Córdoba thought San Joaquín had been poorly constructed in the first place, and enclosed a plan to illustrate his meaning.[45]

Just as the fortress lacked material defensive strength, it also lacked trained manpower. There were only ten men assigned to the battery, and four of those had no knowledge whatsoever of artillery. Even if all the men in the presidio rushed to San Joaquín in time of attack, Córdoba lamented, there would not be enough to man the thirteen inadequate pieces of artillery. Thirty-eight cavalrymen were permanently assigned to escort duty away from the

---

[45] This plan, and others of Monterey that Córdoba mentioned in this report, have not been identified or located in the archives.

port and could not be expected to aid in case of attack.

Córdoba pointed out that not only did flabby defenses make the harbor of San Francisco an easy target for foreign attack, but the entire coastline to Monterey was open and offered several good anchorages for the enemy at Punta de Almejas, Rancho de los Padres, and Santa Cruz. Córdoba's simple remedy for this weakness was an increase in the cavalry complement, which could quickly rush to any minor harbor threatened by invasion. At San Joaquín, Córdoba despaired of ever being able to transform it into a proper fortification because of laborer and materials shortages necessary for modifications. He also thought that another battery at Punta San Carlos across the bay, in present Marin County, was essential for adequate defense. First, though, a reconnaissance of the rocky bluffs would be necessary, and building supplies would have to be ferried to the site because the long land distance around the immense bay left no alternative. He despaired of the feasibility of ever establishing a battery at San Carlos.

The situation at Monterey was no better. The existing battery of ten cannon, according to Córdoba, could only hit an enemy ship if it deliberately placed itself dead in front of the artillery. The guns were short range and consequently useless since enemy ships could easily anchor beyond fire range and still be at safe port. The critical situation in San Francisco and Monterey caused Córdoba to conclude that only excessive expenditure could relieve the present incapacity to defend. Therefore, he suggested that several ships should patrol the coast constantly and thereby provide a mobile battery arrangement to defend the California settlements. Costansó had come to the same conclusion just a few years earlier in contemplating the gigantic expense requisite to refurbish adequately California harbor defenses. Though Córdoba had landed at San Diego, he had not reconnoitered defense installations and reported only that he had heard a wooden platform mounted with four cannon served as the sole defense. He promised to report more after seeing for himself.

Several months later, Córdoba had written to Borica that San Diego, like the other California ports, was incapable of proper defense. Without huge expenditures, the engineer said, it was impossible to build fortifications that could safely guard against naval attack. Córdoba recommended an increase in troops, counting on land defense if an enemy should attack.[46] The impossible defense situation at San Francisco had been swallowed and had not set well at all with the viceroy. After reading Córdoba's report of September 1796, he informed Borica that something had to be done immediately at California's most important port. In April of 1797, the governor accordingly notified Córdoba, in Monterey at the time, that as soon as he had completed the battlements in the capital he should return to San Francisco. The plan was to transfer some artillery from San Diego or some other port where it was not so vital, to San Francisco. A second battery was to be constructed on Yerba Buena Island, at the point that could best protect against an enemy ship anchoring in the bay. The governor realized that men were in short

---

46 Borica to Córdoba, Monterey, Jan. 20, 1797, BL, *Calif. Arch.,* 14.

supply for a new battery, but urged Córdoba to do his best regardless.[47]

Córdoba was unable to work on the modifications of San Joaquín and construction of Yerba Buena full time because at that particular moment, spring of 1797, he became deeply involved in his most ambitious California project, the founding of Branciforte. In fact, activities concerning the new town took up most of the engineer's time in California. This is an indication of the importance placed on the new town by the governor and viceroy, an attitude which seems unusual considering that at the same time, they were terribly worried about the defense situation. Perhaps it is with hindsight, knowing that Branciforte was doomed as a failure because of its proximity to the mission of Santa Cruz, that it seems most inappropriate that the valuable engineer was dallying about with a new town when he might have been working full time on a viable defense arrangement. But the viceroy was willing to finance and encourage a new town which, after all, was named in his honor; whereas a new battery, effectiveness of which could only be doubtful, was named for the pungent mint that grew wild on its rocky crags.

Only six months after Córdoba's arrival in Alta California, the governor ordered the engineer to the mission of Santa Cruz to begin search for a good townsite.[48] The viceroy had ordered a new town built to provide quarters for the new Catalonian Volunteers under Alberni. Branciforte had provided Córdoba with a report by the royal tribunal of accounts

[47] Borica to Córdoba, Monterey, Apr. 4 and May 3, 1797, BL, *Calif. Arch.,* 14.

[48] Borica to Córdoba, Monterey, June 18, 1796, BL, *Calif. Arch.,* 14.

concerning establishment of the new town.[49] Together with Borica, Córdoba, reconnoitered the area from Santa Cruz to Arroyo de los Pájaros and, with Alberni, the areas of La Alameda and the mission of San Francisco. Córdoba was to make a report to the king on the best location for the new town and all the pertinent information relating to its establishment.[50]

Córdoba's report was ready in July of 1796.[51] He stated without question, that of the three areas suggested for the new town, only the one near mission Santa Cruz was suitable. The San Francisco area was bad, principally because of the sandy soil, strong winds, and lack of fresh water. The area called La Alameda was not suitable for irrigation and lacked building materials nearby. The site near Santa Cruz, called Arroyo de los Pájaros, however, had everything. The soil was fertile, easily adaptable to irrigation for crops, and blessed with good pasture for livestock. Wood, stone, lime, dirt for adobe and bricks, and other building materials were all available in abundance. The sea was nearby, which of course would provide fish for food as well as easy shipping facilities. There was enough good land to support the new town and still not infringe upon the Indians at the mission. Being thirty leagues from San Francisco and 25 from Monterey put the new town at an ideal spot for communications, and for supplying

---

[49] "Informe del Real Tribunal y Audiencia de la Contaduría Mayor de Cuentas sobre fundación de un pueblo que se llamará Branciforte," Mexico, Nov. 18, 1795, BL, *Calif. Arch.*, 7. Paragraph 8 of this report specifies that Córdoba, working with the governor and soldiers, was to found the new town.

[50] Borica to Córdoba, Monterey, June 16, 1796, BL, *Calif. Arch.*, 52.

[51] "Informe acerca del sitio de Branciforte," Presidio de San Francisco, July 20, 1796, BL, *Calif. Arch.*, 52. See Appendix F.

each with the excess produce and livestock that Branciforte was sure to yield. All in all, Córdoba was optimistic about the townsite he had chosen. He was sure the new town would be a big success within a short time.

By May of 1797, Borica had received word from Viceroy Branciforte that Córdoba's site recommendation was approved.[52] In informing the engineer of this, Borica also gave instructions to Córdoba about the next step in establishing the new town. As soon as Córdoba completed his work in San Francisco, he was to return to Santa Cruz, and from there begin laying out Branciforte, prepare maps and plans of proposed buildings, irrigation projects, and fields; and also prepare a cost estimate for the entire operation. The governor explained to Córdoba that Viceroy Branciforte had no intention of allowing his town to become another Los Angeles or San José. Those two places, after fifteen and twenty years, respectively, had miserable, squalid huts instead of proper buildings and houses. The viceroy wanted Branciforte to be complete from the beginning, a rather over-ambitious desire considering sparse California settlement and supplies.

By August, Córdoba reported to Borica that he had staked out the new town and had constructed temporary houses and arranged a water supply so that major construction and farming could begin.[53] The engineer was busy gathering building materials and laying out fields for cultivation. Only a week later, construction had to be suspended during the rains,

[52] Borica to Córdoba, Monterey, May 15, 1797, BL, *Calif. Arch.,* 14.
[53] Córdoba to Borica, Santa Cruz, Aug. 12, 1797, BL, *Calif. Arch.,* 10.

but people were already beginning to work on the lands assigned to them for crops.[54] In the following months, Córdoba continued to direct the construction of Branciforte, having supreme power over soldiers and Indians working at the site.[55] The engineer made plans of the villa and surrounding area, and also, finally, had time to complete the long-awaited map of Alta California as well as plans of the new batteries at Yerba Buena, Santa Barbara, and San Diego.[56]

In October of 1797, funds for construction at Branciforte ran out and the governor ordered Córdoba to return to San Francisco after leaving instructions for completion of work in progress with Franco de Molino from Mission Santa Cruz.[57] The engineer's major work had been completed. Orders for Córdoba's return to Mexico had already been dispatched in the fall of 1797, and the remainder of Córdoba's time in California was busy, but really only an epilogue to his major purpose in the far north.[58] Word was received in April of 1798 that Córdoba had been ordered back to Mexico.[59] Up until that time and afterwards for several months, until Córdoba embarked to sail south, he was occupied with the recon-

---

54 Borica to Córdoba, Monterey, Aug. 21, 1797, BL, *Calif. Arch.*, 14.

55 Gabriel Moraga to Borica, Santa Cruz, May 29, 1797, BL, *Calif. Arch.*, 7.

56 Borica to Córdoba, Monterey, Sept. 30 and Nov. 26, 1797, BL, *Calif. Arch.*, 14. Of these maps, only Córdoba's projected plan of the villa of Branciforte has been located. A brief description of Córdoba's connection with Branciforte and the plan are to be found in Lesley Byrd Simpson, *An Early Ghost Town of California: Branciforte. See also* illustration on page 137, herein.

57 Borica to Córdoba, Monterey, Oct. 24, 1797, BL, *Calif. Arch.*, 14.

58 Branciforte to Borica, Orizaba, Nov. 27, 1797, BL, *Calif. Arch.* 14.

59 Borica to Córdoba, Monterey, Apr. 7, 1798, BL, *Calif. Arch.*, 14.

naissance of the coast between San Francisco and Santa Cruz, in search of good sites for new settlements.[60] Córdoba had found time to finish repairs on San Joaquín, where he had improved firepower by placing six cannon to defend the entrance to the bay and three on each flank. Yerba Buena too was made a working battery capable of defending the land between it and San Joaquín, or the point of the peninsula; and also the channel between Alcatraz and Punta de los Méganos.[61] The engineer had also managed to clarify the border between Santa Clara mission and the town of San José.[62] Córdoba taught the Indians of Santa Clara how to build houses in their area and instructed others how to maintain the dam he had constructed near Branciforte.[63]

On October 11, 1798, Governor Borica gave Córdoba a boarding pass addressed to Captain Juan Matute for return to Mexico.[64] By the time Córdoba left California, he had indeed completed quite a number of engineering commissions. He had made reports and drawn maps of Santa Barbara, San Francisco, San Diego, Monterey, and the whole of Alta California. He had fortified San Joaquín to a greater strength than ever before and built a supplementary battery to guard the all-important San Francisco Bay. Most outstanding, at least at the time, he had

---

[60] Borica to Córdoba, Monterey, Mar. 3 and July 20, 1798, BL, *Calif. Arch.*, 14.

[61] Borica to Córdoba, Monterey, July 2 and Aug. 3, 1797, BL, *Calif. Arch.*, 14.

[62] Borica to Córdoba, Monterey, Aug. 8, 1797, BL, *Calif. Arch.*, 14.

[63] Borica to Córdoba, Monterey, Mar. 15, and May 3, 1798, BL, *Calif. Arch.*, 14.

[64] Borica to Juan Matute, Monterey, Oct. 11, 1798, BL, *Calif. Arch.*, 14.

chosen the site for, and founded the new town of Branciforte.[65] Agreeing with Costansó, Córdoba knew that his defense improvements were really only stop-gap measures put into operation instead of vastly more expensive plans that would have provided certain security for California. Like Costansó also, Córdoba believed that an increased population was the real key to success for California. Though they never met in California, the scene of major labors of each, Costansó and Córdoba undoubtedly agreed on the whole California question. Brothers in the Corps and also in defense principle, the two engineers acted together, albeit apart, for strengthening the northern outpost with limited resources.

By the time aid was sent to California and Córdoba assisted in defense there, Costansó had very definitely contributed to the welfare of the western Borderlands. His intimate association with California from its founding through the period of major attention from officials in Mexico City and Spain qualifies him as one of the most important participants in Spanish California. The brief time he spent in Sonora and Lower California helped to establish his reputation as a cartographer and provided data used by other mapmakers later. And if Costansó's Borderlands associations were not enough to establish him as an excellent engineer, his numerous accomplishments in the capital earned additional accolades for him and the Corps.

After his return from California in 1770, Costansó was put in charge of construction of the Hospital of

---

[65] AGS, *Guerra Moderna*, 7243; and AHN, *Estado*, 4290.

San Andrés,[66] and then of the new mint, or *Casa de Moneda,* a project that occupied him more or less steadily for eight years.[67] In 1779, he began work on the new gunpowder factory at San Miguel de Perote, site of new defenses to guard the ascent to Mexico City from Veracruz.[68] On October 22, 1779, Viceroy Martín de Mayorga authorized establishment of the Academy of San Carlos for the fine arts; and put Miguel Costansó in charge of design and construction. Later, the engineer served as a geometry professor and a director at San Carlos.[69] Continuing as a consultant and designer of the many new public service buildings that graced Mexico during this period, Costansó wrote a report on construction of the *Casa de Recogidas,* an institution for the rehabilitation of prostitutes and a home for vagrants, in 1781.[70] On November 19, 1784, fire destroyed the old powder factory at Chapultepec, and Costansó was charged to produce reconstruction plans.[71] Costansó supervised a street paving project for the capital, in which he introduced a new method of slab paving; and he provided a plan for supplying water to the city from Lake Chalco in 1783.[72] Costansó submitted a cost schedule for improvement and beautification of the *Plaza del Volador,* near the Zócalo, in 1791;

---

[66] AGN, *Virreyes,* 142.

[67] Service record, Veracruz, Mar. 20, 1784, AGS, *Guerra Moderna,* 7241.

[68] *Ibid.*

[69] AGN, *Casa de Moneda,* 229; and Miguel Costanzó (sic), "Elementos de Geometría," 1785, ms. copy in Ariz. St. Univ. Spec. Coll., Porrúa Collection.

[70] Miguel Costansó to Francisco Romá y Rosell, Mexico, Jan. 20, 1781, AGN, *Obras Públicas,* 5.

[71] AGS, *Guerra Moderna,* 7241.

[72] AGN, *Virreyes,* 142; *Ayuntamiento,* 202; and *Obras Públicas,* 6.

then two years later, drew plans for fountains for the same square.[73] Just a year later, Costansó was working on modernization and new ornamentation of the *Plaza Mayor,* which included enhancement of the facade of the viceregal palace.[74]

Costansó was constantly utilizing his drafting and cartographical skills in the production and copying of maps and plans. For example, between 1790 and 1794, the engineer made 224 maps by order of Viceroy Revilla Gigedo.[75] In 1797, Costansó wrote a special report, accompanied by his own map, of the Orizaba area, relating to improvement of the road and its defenses to the capital from Veracruz.[76]

Obviously, Costansó had a very active career. His long years of service and his meritorious conduct and talents provided him with regular promotions in rank and class. On October 2, 1795, he became full colonel and chief engineer simultaneously.[77] Seven years later he became brigadier in the army and was already a director and top-ranking corpsman in New Spain. In 1813, he achieved the rank of field marshal,[78] and by then his class in the Corps was that of sub-inspector, the top post under the engineer general created by the Ordinance of 1803. By 1809, Costansó was the fourth ranking member of the entire Corps.[79]

[73] AGN, *Provincias Internas,* 121; and *Obras Públicas,* 36.

[74] AGN, *Obras Públicas,* 36.

[75] Bill presented to Revilla Gigedo by Costansó, Mexico, Feb. 12, 1794, AGN, *Ramo Civil,* 1408.

[76] AGS, *Guerra Moderna,* 7244.

[77] Service record, 1795, AGS, *Guerra Moderna,* 3794.

[78] AGM, *Expediente personal de Dn. Miguel Costansó.*

[79] "Lista general de los oficiales que componen el Real Cuerpo de Ingenieros . . . 1809," *Memorial de Ingenieros del Ejército,* May 1908, pp. 347-51.

Throughout his 52 years of service in the Corps of Engineers, Costansó consistently received the highest possible ratings on yearly service reports by his superiors. At various times Costansó's superiors added appraisals of their own not required on official forms. Viceroys Croix, Bucareli, Mayorga, Gálvez (both father and son), and Flores, all praised Costansó and trusted him with the most important commissions.[80] Manuel Santistevan, Costansó's superior officer in 1775, rated him as intelligent and dedicated. By 1784, Miguel del Corral, who was Costansó's immediate superior for many years, commented that Costansó's conduct was above reproach, his drawing was outstanding, that he had great talents in many fields; and that he was capable of any commission assigned to him.[81]

Little is known of Costansó's personal life. In 1776, at the age of 35, he petitioned for permission to marry, in keeping with regulations of the Corps. Authorization came about a year later, and he married well, for his wife was from one of the best families of New Spain. Manuela de Aso y Otal's father was provincial governor of the Marquesado del Valle, a hereditary holding in the family of Hernán Cortés.[82] Costansó was held in high regard in the capital. He served under seventeen viceroys of New Spain and was a trusted adviser on civic im-

[80] Manuel Antonio Flores to Marqués de Sonora, Mexico, Sept. 27, 1787, AGN, *Virreyes*, 142.

[81] AGN, *Historia*, 568; AGS, *Guerra Moderna*, 7241; and AGI, *Audiencia de Mexico*, 2472.

[82] AGM, *Expediente de casamiento*, with *Expediente personal, 1813;* and AGN, *Historia*, 568.

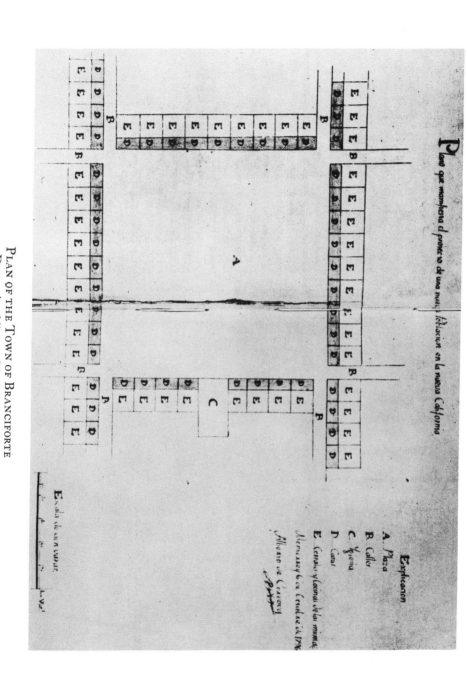

PLAN OF THE TOWN OF BRANCIFORTE
Drawn by Alberto Córdoba in 1796.
See page 131. From Lesley Byrd Simpson, *Branciforte*

provement and beautification of the capital,[83] as well as California and other defense problems.[84] Within the Corps he was appreciated and commended as a model representative of the high standards set for engineers.

A half-century of distinguished and illustrious service in New Spain ended for Miguel Costansó with his death, apparently from natural causes, in Mexico City on September 27, 1814.[85] During his time, the activities of the Corps had expanded and extended to their greatest glory. Though his career is certainly the most impressive of the military engineers who traveled and worked in the Borderlands, it still is only part of the story. Just as others had trod the lands of the Southwest before his days in Sonora and the Californias, so too, were others to march after him. While Costansó's Borderlands career was most closely tied to the founding and continuing colonization of Alta California, other corpsmen labored in the northern interior reaches of New Spain. The Spanish government and royal engineers pursued a new reform policy in the north where defense was not yet secure.

---

[83] Ernesto de la Torre, ed., *Instrucción Reservada qué dió el Virrey don Miguel de Azanza a su sucesor don Félix de Marquina*, p. 96; José Bravo Ugarte, ed., *Instrucción Reservada que el Conde de Revilla Gigedo dio a su sucesor en el mando, Marqués de Branciforte*, p. 176.

[84] Many references to Costansó's *Narrative* and other writings exist in the archives, all lauding the engineer's scientific precision and descriptive talents. Among them are three undated California manuscripts, all of which cite Costansó as an authority, and quote and praise his work, MN, mss. 317 (Papeles Varios IV), 567 (Virreinato de Mexico I), and 621 (Descripción de California).

[85] Joachim Blake to Secretaría del Estado y del Despacho Universal de Indias, June 3, 1815, AGM, *Expediente personal, 1813*.

5

The Commandancy General
of the Interior Provinces

Recommendations that the Marqués de Rubí and engineer Nicolás de Lafora made after their review of the frontier presidios had led to the Regulations of 1772. Though Hugo Oconor, as Commandant-Inspector, had been sent to put into effect the new orders, and though he labored assiduously at his task, officials in Mexico and Spain still were not satisfied with the state of defense on the frontier, especially in the western areas, where the Apaches were more troublesome than ever. Accordingly, the Commandancy General of the Interior Provinces was created in 1776 as a potentially more effective organization for dealing with both Indians and foreign menaces to Spain's northernmost colonial provinces.[1]

Brigadier Teodoro de Croix, nephew of the Mar-

---

[1] For explanation of factors leading to creation of the Commandancy General, and varying interpretations, see Lillian E. Fisher, *Viceregal Administration in the Spanish American Colonies*, p. 275; Chapman, *Spanish California*, p. 307 and 386-416; Bobb, *Viceregency of Bucareli*, p. 144; and Alfred Barnaby Thomas, *Teodoro de Croix and the Northern Frontier of New Spain, 1776-1783*, pp. 16-17.

qués de Croix, former viceroy of New Spain, was appointed by Charles III as first commandant general. He arrived in Mexico in January of 1777, carrying instructions from the king on his duties and aims in the new post.[2] Croix remained for a time in the vice-regal capital familiarizing himself with his duties and from available information on his command, making plans for the future. Croix must have had many things on his mind, for he was to possess powers that rivaled those of the viceroy. Besides seeing to strictly defensive matters, the commandant general was to establish a capital at Arispe, Sonora, and he was to encourage the founding of new civil settlements. His jurisdiction included both Californias, Sonora, Sinaloa, New Mexico, and the eastern border regions. He had a great deal of work to do and he knew it.

One of Croix's foremost requests for aid from the king was that two army engineers be assigned to his command.[3] In a letter to Minister of the Indies José de Gálvez, Croix pointed out that although Rivera and Rubí both had been accompanied by engineers, a completely accurate map of the northern frontier still was not available. In addition to mapping, he also required engineers for construction of the *casa de moneda,* or mint, in Arispe. Gálvez understood the value of engineers, and had great respect for the Corps. Upon receipt of Croix's request, he imme-

---

[2] AGI, *Audiencia de Guadalajara,* 242 and 390; AGS, *Guerra Moderna,* 7049.

[3] Teodoro de Croix to José de Gálvez, Mexico, Mar. 24, 1777, AGI, *Audiencia de Guadalajara,* 516.

diately directed the Minister of War, the Conde de
Ricla, to arrange for selection of two ordinary engineers with appropriate qualifications.[4] A week after
the three chiefs of the Corps had received the request
for engineers to help Croix, one of them had reviewed the records and proposed two men.[5] In the
following months the other chiefs concurred in the
choice of Manuel Mascaró and Gerónimo de la
Rocha for the tasks of drawing maps and supervising
construction of the *casa de moneda* at Arispe.[6] Ricla
informed Gálvez of the decision. The Minister of
the Indies replied that the engineers should proceed
immediately to Cádiz for embarkation. Ricla prepared transfer orders for Mascaró in Cartagena and
Rocha in Barcelona. Within two weeks, royal orders
and passports were awaiting the engineers at the port
of Cádiz.[7] Rocha and Mascaró each were to receive
800 pesos annual salary to begin on the day of embarkation.

Croix's request for two engineers had received special, speedy attention from José de Gálvez, undoubtedly because the job was thought to be especially
important. The chiefs of the Corps and the Minister
of War had acted quickly since Gálvez stressed the

4 Gálvez to Conde de Ricla, the Palace, July 15, 1777, AGI, *Audiencia de Guadalajara,* 516.

5 Ricla to Abarca, Sabatini, and Lucuce, the Palace, July 20, 1777; and Sabatini to Ricla, Madrid, July 28, 1777, AGS, *Guerra Moderna,* 3066.

6 Abarca, Sabatini, and Lucuce to Ricla, Madrid, Sept. 6, and 18, 1777, AGS, *Guerra Moderna,* 3066.

7 Ricla to Gálvez, San Ildefonso, Sept. 18; Gálvez to Ricla, Sept. 21; Ricla to Gálvez, Sept. 22, 1777; royal orders, San Ildefonso, Sept. 28, 1777; and Francisco Manxon to Gálvez, Cádiz, Oct. 7, 1777, AGI, *Audiencia de Guadalajara,* 516.

urgency of the matter. In July of 1777, when Gálvez was already arranging for the requested engineers, Croix, anxious to get started, took it upon himself to find an engineer. He requested the viceroy to assign to him Carlos Duparquet, whom Croix had known previously and who was then stationed in Veracruz. When Gálvez heard that Bucareli had acceded to Croix's request and had sent for Duparquet, he advised that the two engineers originally requested already had been dispatched and that one should remain under the viceroy's orders in Mexico. Croix had suggested that his engineers be paid as Lafora had been, but Gálvez tersely pointed out that the commissions were different in every respect and that Lafora had been in the Interior Provinces only a short time.[8]

But before Croix had received Gálvez' judgment on this matter, the commandant general started the march north, and took Duparquet with him. The engineer drew a sloppy map of their itinerary[9] that included several serious mistakes. Croix thought Duparquet had been negligent in taking astronomical observations, and apparently he dismissed the engineer. Duparquet was back in Veracruz by the end of 1779, and did not return to the Interior Provinces.[10]

---

[8] Croix to Gálvez, Mexico, July 26, 1777, and Gálvez to Croix, Dec. 21, 1777, AGI, *Audiencia de Guadalajara,* 516.

[9] "Mapa desde Veracruz a los Presidios del Norte y de estos a Chihuahua levantado de orden del Sr. Comandante de Croix por el capitán e ingeniero ordinario Don Carlos Duparquet, desde principio de agosto de 1777 hasta 14 de marzo de 1778," Chihuahua, May 12, 1778, AGI, *Audiencia de Mexico,* 539; and reproduced in Navarro García, *Don José de Gálvez,* following p. 176.

[10] AGS, *Guerra Moderna,* 3794; and BCM, vol. 56.

To replace Duparquet, Croix employed the services
of Luis Bertucat, a volunteer engineer, not a member
of the Corps.[11] Bertucat drew a fine map of Croix's
itinerary from Durango to Chihuahua for the commandant general.[12] He spent a total of six months in
the Interior Provinces, in which time he invented
and manufactured armor plate, which Croix gave for
use to members of his escort.[13]

The two engineers Croix had originally requested
from Corps officials in Spain arrived in Sonora in
November of 1779.[14] The commandant general kept
them busy in the next years with the specific tasks he
had listed in his 1777 request and with even more
projects. Rocha and Mascaró served their chief, the
Commandancy General, and the crown skillfully, contributing many noteworthy services toward strengthening the northern defense system.

Gerónimo de la Rocha y Figueroa was born in
1750, in Orán, the Spanish fortification colony in
North Africa. He served as a cadet in the fixed regiment of Orán, of which his father was commandant;
he studied mathematics in the Academy there. Rocha
took entrance examinations in Barcelona and entered

11 AGS, *Guerra Moderna*, 3794; and Jack D. Holmes, *Honor and Fidelity: the Louisiana Infantry Regiment*, p. 95.

12 "Mapa del Derrotero que hizo el Comte. Gral. Cavro. de Croix por
las Provincias de su cargo desde la Ciudad de Durango hasta la Villa de
Chihuahua, formado sobre las Longitudes del Ingenro. Dn. Miguel Costansó y las Latitudes de dn. Nicolas Lafora en el Año 1778," Chihuahua,
1778, AGI, *Audiencia de Mexico*, 538; and reproduced in Navarro García,
Don José de Gálvez, following p. 176.

13 AGI, *Audiencia de Guadalajara*, 271. See Navarro García, *Don José
de Gálvez*, p. 405, for consequences of Bertucat's invention.

14 AGS, *Guerra Moderna*, 7241.

the Corps of Engineers on September 4, 1771, as second lieutenant and assistant engineer. He served at the fortification of San Fernando de Figueras, with an occasional visit to his home in Orán, until 1777, when he was transferred to the Interior Provinces of New Spain. Shortly before going to America, Rocha was promoted to lieutenant and extraordinary engineer.[15]

Once arrived on the northern frontier, Rocha became involved in two distinct activities. First of all, he served as engineer to Jacobo Ugarte y Loyola,[16] newly appointed military governor of Sonora and later commandant general. Croix assigned Ugarte on April 17, 1780, to move the presidios of Sonora back to locations they had occupied before the Rubí recommendations and subsequent Oconor changes. Croix had studied this problem of presidio alignment while still in Mexico and had received the king's approval in 1778 to readjust the line. Ugarte's job was to utilize Rocha's expert advice in scrupulously examining the new and old sites.[17]

Rocha kept diaries of his travels with Ugarte, which lasted throughout the spring and summer of 1780. In February and March Rocha had transferred Santa Cruz de Terrenate back to the old site of Las Nutrias, a change of about twelve miles. From April through July, Ugarte and Rocha tramped through Sonora, supervising the presidio transfers.[18] All along the way, Rocha must have been taking astronomical

---

[15] Service record, May 6, 1777, AGS, *Guerra Moderna*, 3793.

[16] Thomas, *Teodoro de Croix*, p. 51.

[17] Thomas, *Teodoro de Croix*, pp. 148-49 (Paragraphs 299 and 300 of Croix's "Report of 1781").

[18] Rocha's diaries and other papers are located in the Harvard Col. Lby., Sparks Coll., ms. 98, vol. VII.

observations and making notes for maps that he drafted a short time later.

On March 6, 1781, Commandant General Croix appointed Rocha as interim captain of the presidio of Fronteras and commander of its cavalry company. Croix also ordered Rocha to supervise construction of new facilities at Fronteras, and neighboring Terrenate. Rocha reported to Croix nineteen months later that shortly after taking command at Fronteras, he had begun to realize that it would be impossible to comply with his orders. Required building materials, workers, and craftsmen were to be found neither in the presidios, nor anywhere else in the vicinity, although Rocha reportedly did his best to find the supplies and personnel he required. Rocha said that the few masons there preferred to work in Arispe rather than in his district because they were afraid of "bad" climate around Terrenate.[19] Rocha recommended that the presidio be moved once again, five or six miles either up- or downstream from Las Nutrias. He reiterated what he had reported to the commandant general some months earlier: it simply was not possible to build new facilities without more men to serve as a guard, and additional men to escort supply trains through that Apache-infested territory.

Rocha lamented that it was absolutely impossible

---

[19] The main reason Rubí had given for transferring Terrenate from Las Nutrias to the new location, a site called Quiburi, was that the former site had bad water and an unhealthful climate, making it necessary for settlers to do their farming some distance away, on the banks of the San Pedro. Croix had ordered the presidio restored to marshy Las Nutrias. Therefore Rocha's comment about bad climate is undoubtedly correct. The Quiburi site, near present Fairbank, Arizona, has been excavated and studied by archæologists in recent years. See Charles C. DiPeso, *The Sobaipuri Indians of the Upper San Pedro River Valley, Southeastern Arizona*, pp. 55-87.

to get laborers; even Croix, with his prestige and power, would not be able to persuade men to work at damp and dangerous Las Nutrias and at Fronteras. The engineer dejectedly stated that he just might as well give up the project as a failure. His own health was broken; he had begun to suffer when he first arrived at Fronteras, and punitive forays he had led had only weakened him more without producing any substantial results in the Indian war. He was bedridden at the time he wrote to Croix and pleaded for any other assignment, which he said he would gladly perform as soon as he was well. He even promised to return to the presidio to direct labor on the wall surrounding the fort if Croix could arrange to make workers available. Conscientious servant of the crown that he was, Rocha closed his forlorn letter by apologizing, ashamed that he was unable to complete his commission.[20]

Croix must have had great faith in Rocha. He had requested promotion to captain for the engineer, which was conceded on February 20, 1782. The commandant general wrote to Gálvez explaining Rocha's illness and other problems, and reported that he had allowed Rocha to return to Arispe to recuperate. Croix replaced him at Fronteras with the former first lieutenant at Santa Fé, Manuel Azuela,[21] who was to remain in Sonora as an Indian fighter for many years.

In Arispe, Rocha prepared maps and plans and probably helped his fellow engineer with construc-

---

[20] Rocha to Croix, Fronteras, Oct. 7, 1782, AGI, *Audiencia de Guadalajara,* 518.

[21] Croix to Gálvez, Arispe, Dec. 30, 1782, AGI, *Audiencia de Guadalajara,* 518.

tion of buildings in the new capital. In 1780, Rocha had drawn an impressive map of northern Sonora,[22] the most detailed map of the area produced during the Spanish period.[23] The map, specifically requested by Croix,[24] shows the itinerary of the review made by Ugarte and Rocha that the commandant general had ordered in 1780. It locates Apache strongholds south of the Gila River and east of the San Pedro within Spanish occupied territories, and shows the location of other local Indian groups. Relocated presidio sites are distinguished from the old places by color. Rocha's own suggestion for location of the presidio cordon is also included. Mountains and river courses are stylized, but an attempt was made to indicate gradations of height and true flow.

Following his stint as captain at Fronteras, Rocha produced another map that incorporated some of the practical knowledge he had acquired about fighting Indians.[25] A very detailed and excellent map, it shows the terrain for a projected expedition against the Gila Apaches. Portions of Sonora and Nueva Vizcaya encompassing the Gila and San Pedro River valleys are shown very clearly, with presidios and former sites, pueblos, ranches, mining settlements,

---

22 "Mapa de la Frontera de Sonora para el Establecimto. de la Linea de Presidios," Arispe, Sept. 4, 1780, in British Mus. Mss. Room Add. 17661a; and reproduced in Navarro García, *Don José de Gálvez,* following p. 456. *See also* illustrations on pp. 155-56, herein.

23 Navarro García, *Don José de Gálvez,* p. 456.

24 Thomas, *Teodoro de Croix,* p. 149 (Paragraph 302 of Croix's "Report of 1781").

25 "Mapa del Terreno que ha de vatir la Expedion. que deve executse. contra los Apaches Gileños," Arispe, Feb. 20, 1784, AGI, *Audiencia de Mexico,* 586 (orig.) and 541; and reproduced in Navarro García, *Don José de Gálvez,* following p. 288.

and watering places. Mountains are completely stylized but well placed, except around the edges of the
map where Rocha's accuracy faltered through lack
of personal reconnaissance. In the left margin of the
map Rocha wrote an explanation of the map and
campaign plan.

In 1785, recovering from his illness in Arispe,
Rocha spent some time in New Mexico,[26] where he
may have aided Governor Juan Bautista de Anza in
campaigns against the Moquis, Utes, Apaches, and
Comanches. By August of 1787, he was back in
Spain;[27] his tour of duty in America was complete.
For a while he worked in Andalucía and then was
transferred to his native Orán and from there to the
coast of Granada. In 1790, while on leave in Madrid,
Rocha helped extinguish the fire in the *Plaza Mayor,*
for which he was commended. In April of 1794,
Rocha was assigned to the Catalonian campaign of
the war France had declared in 1793 against Spain.
There, promoted to lieutenant colonel, he was soon
put in charge of engineering particulars at the fortification in Lérida.[28] After the peace, Rocha was transferred to Catalonia where he was serving in 1796.
This is the last available mention of the engineer.

Rocha's companion serving in the Interior Provinces was Manuel Augustín Mascaró, born in 1747
in Barcelona. Like so many other members of the
Corps, Mascaró entered military service very young.
He was a cadet in the Royal Spanish Guards for five
years during which time he studied mathematics at

[26] AGS, *Guerra Moderna,* 3794.
[27] AGS, *Guerra Moderna,* 5837.
[28] AGS, *Guerra Moderna,* 3794.

the Academy in Barcelona. On September 26, 1769, he was admitted to the Corps as second lieutenant and assistant engineer. He served in his native Catalonia, Orán, and Cartagena before being called to assist Croix in northern New Spain. Promoted to lieutenant and extraordinary, he sailed for America in April of 1778, reaching Arispe on November 13, 1779.[29]

While Rocha excelled in military cartography and fortification construction, Mascaró was concerned primarily with civil architecture and engineering. He began his close association with the commandant general even before arriving in the capital of the Interior Provinces. Coming north from Mexico City, Mascaró paused at Chihuahua and from there proceeded on to Arispe accompanying Croix, keeping a diary of their trip. In this daily journal,[30] which covers the time from September 30 to November 13, 1779, Mascaró showed the engineers' customary precision, completeness, and eye for detail. The diary is not greatly different from other travel logs of the same period except that it is more ordered, more exact, and more scientifically oriented than those produced by persons without engineering training and outlook. Mascaró listed every change in direction of the road, each junction, the time and distances between each spot and the next. He noted the taking of astronomical observations, where watering places were to be found, and the location of surrounding mountains. This route had been traveled many times

29 Service record, Jan. 1, 1787, AGS, *Guerra Moderna*, 5837; and AGS, *Guerra Moderna*, 7241.

30 "Diario del Yngeniero dn Manuel Mascaró desde la Villa de Chihuahua ál Pueblo de Arispe, en la Pimeria alta Gobernacion de Sonora, año de 1779," BL, microfilm of *Mexican Mss.* 400-02, ms. 57, 326.

and was well known to at least some of the Indian guides, the picket of dragoons, and the company of cuirassiers that made the journey with Croix and Mascaró. The engineer's copious detail can only be ascribed to thorough training and a conscientious attitude. In new territory, Mascaró must have realized that every geographical point he noticed upon his entrance might help him later in the commissions Croix would give him.

Once in Arispe, Mascaró remained there except for infrequent trips to nearby places in Sonora. One such expedition was ordered by Croix in 1779. Mascaró traveled down the Sonora River valley to the area of Pitic to plan construction for the presidio and an irrigation ditch for adjacent fields. Mascaró completed his project to Croix's satisfaction, and the engineer's plans and reports were forwarded to the auditor of war to explain the recently effected site change of the presidio of Horcasitas to Pitic. Croix reported that he had funds available and had ordered Mascaró's work plans put into operation.[31]

Except for occasional reconnoitering activities such as the one to Pitic, Mascaró lived in Arispe for five years after his arrival in Sonora and worked mostly in the provincial capital.[32] In 1781, he wrote an impressive description of the new capital and its environs as a background study to converting the squalid mission village into a town worthy of the name capital of all the Interior Provinces.[33] Mas-

---

[31] Thomas, *Teodoro de Croix*, pp. 216-17 (Paragraphs 512-513 of Croix's "Report of 1781").

[32] AGS, *Guerra Moderna*, 7241.

[33] "Descripción, y actual Estado del Pueblo, y Misiones de Arizpe, que S.M. ha destinado en sus Reales Instrucciones, para Capital de estas Pro-

caró's description of Arispe is far superior to others of its era. Compared to Fersén's work on Sonora and Sinaloa, Mascaró's piece is a marvel of tight, succinct prose and a storehouse of valuable data and opinions on the people and the area. The tone of his report is critical, in both negative and positive ways. For example, he began by describing the attractive physical setting, lay-out, and buildings of Arispe. Yet he was quick to point out that the streets were irregular and crooked, the one irrigation canal was badly constructed and directed, and the road entering town was twisting and bothersome. Mascaró's civil engineer's eye was well attuned to the shortcomings of Arispe.

Mascaró continued with a detailed architectural description of the church and other buildings, including the mission house where the commandant general lived. His most frequent criticisms were that the structures were ill proportioned, usually too low for his tastes; dark and close inside; and had partially or completely unserviceable roofs, either through neglect or initially poor construction. Croix's quarters, which had "no comfort whatsoever," were nonetheless the best in town, and the only one with living space raised off the ground. The secretary's chambers, where Mascaró may have worked, had the distinct disadvantage of opening onto the choir of the church, where one had "to tolerate the annoyance of organ and singers practicing." But aside from the disturbing music during working hours, Mascaró delighted

---

vincias internas, Clima, Producciones, y Calidades de su terreno, Caracter, Govierno Civil, y Militar de sus Habitantes, con una corta noticia de los Proiectos que se han delineado segun lo ha permitido el terreno," 1781, unsigned copies in MN, mss. 485 (Miscelanea) and (two copies in) 567 (Virreinato de Mexico I). See Appendix G.

in the rich interior ornamentation of the church, its altars dedicated to the Assumption of the Virgin Mary, Our Lady of Loreto, and Saint Ignatius Loyola.

Like the commandant's quarters, community housing (used as barracks) and the 130 civilian houses were low, of poorly built adobe, and ugly, Mascaró felt. Condition of the barracks was made more unpleasant for soldiers, because it was also used as a jail since there was no other lockup for occasional prisoners. The one merchant in town had the best structure on the main square. Land immediately adjacent to the lower level of the town was idle, but could easily be converted into rich garden patches if Mascaró were put in charge of a simple irrigation project to protect against seasonal heavy flooding. Ample water was available in the area if only a systematic method of controlling the supply were developed. The land was rich enough to support a sizeable population considering the chronic Apache depredations. In Arispe there were 305 Hispanic people and 337 Opata Indians, and many more natives and mestizos in surrounding areas. The climate was salubrious and the seasons regular. Chronic ailments were unknown except for "the French disease,"[34] which was widespread among the Indians due to their poor diet, indifferent sanitary habits, and lack of medical knowledge.

Despite the Indians' inherent indolence, according to Mascaró, they were able to produce many and varied crops. Wheat, corn, and beans were staple

---

[34] "El Gálico" is the term used by Mascaró. The standard translation of "the French disease" is venereal disease or syphilis.

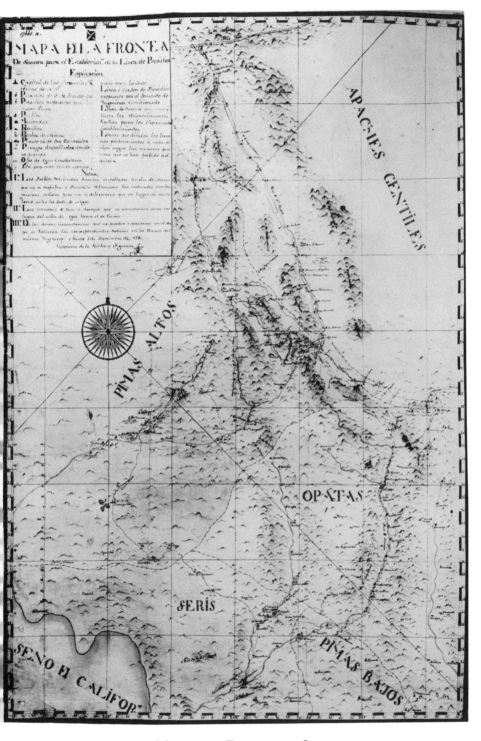

Mapa de la Frontera, 1780
Rocha's Map of the Sonora Defense Frontier.
See page 149. Courtesy, Bancroft Library.

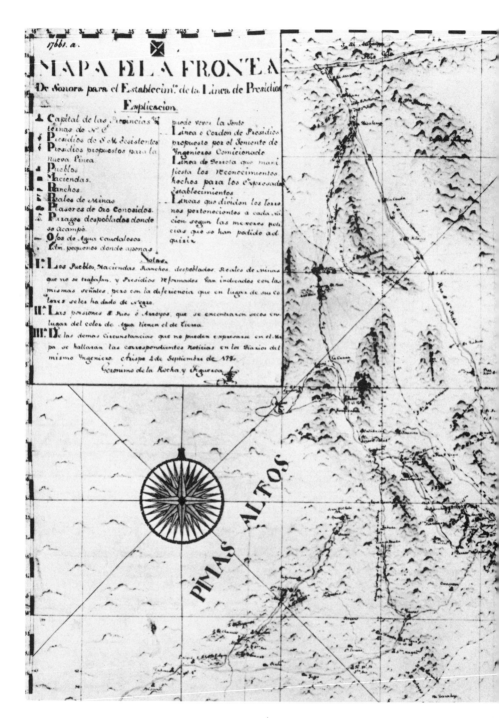

MAPA DE LA FRONTERA
Detail of Rocha's map on the preceding page.

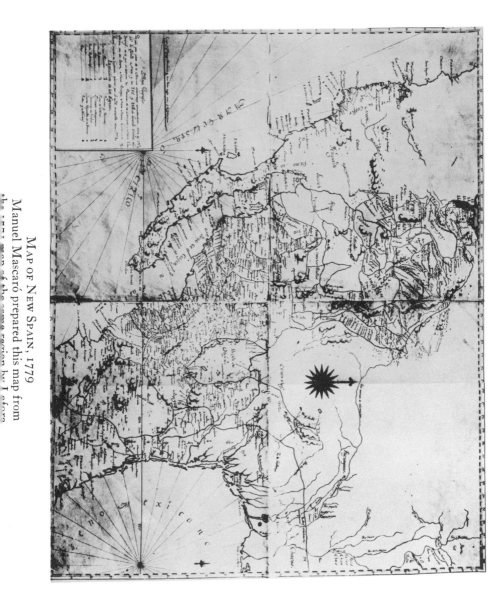

MAP OF NEW SPAIN, 1779
Manuel Mascaró prepared this map from
the 1771 map of the same region by Lafora

products but many varieties of fruit were grown in addition to sugar cane, cotton, chile, and lentils. Mascaró's mouth seemed to water when he wrote that the melons, harvested twice annually, were succulently exquisite; but he lamented the lack of good vegetables. Beef and pork, though, were the best in America, and fish from numerous streams provided variety, and no doubt, an unexpected reminder of Mascaró's native Barcelona. The gourmet engineer went on to list many herbs available in the vicinity, including something similar to European absinthe and others, useful for medicinal purposes as well as for *le haute cuisine.*

Concerning natural resources, Mascaró depicted Arispe to be but one step short of the Garden of Eden. Innumerable types of natural grasses grew in great abundance, as did trees producing woods of every conceivable use. Granite and fine marble and limestone and plaster were available in the hills and mountains for rebuilding the rustic capital. In short, Mascaró said, all the construction materials he would require for his projects could be had at short distance from Arispe. The one exception was timber which, although not too far distant, was difficult to bring down because of a shortage of oxen and inevitable Apache attacks.

Mascaró pointed out another resource of the area, one that was of the utmost interest. In the vicinity were three large gold placers. Some gold was to be found in almost every stream bed in the area, but generally it was too risky for small mining camps because of the Apaches. Other mining sites were numerous: active silver, copper, lead, and iron works, and abandoned gold and silver sites that were either exhausted or in areas much too close to Apache haunts to be worked. Mascaró's discussion of mining activity sounds much the same as Fersén's a decade earlier: the area was rich but the Apaches' constant raids would not allow steady extraction activities.

The remaining portion of Mascaró's 1781 description deals with the character of the Opatas native to the district and the usual political, military, and ecclesiastical organization. Like all other Indians, ethnocentric Mascaró said, the Opatas were indolent, superstitious, untrustworthy, and dirty; but still they were industrious, long-suffering, robust, and resolute. Mascaró thought these redeeming features, their love for the Spaniards and the king, and their ready acceptance of Christianity would make them loyal, obedient, and valuable vassals. Father Eusebio Francisco Kino a century before had seen the same promise in these people.

Mascaró's 1781 description makes better reading than most documents on similar subjects of the time. He was an astute observer and an ambitious planner. He was able to correct at least some of the deficiencies he noted in Arispe. He planned and supervised construction of a dam on the Bacanuchi River for irrigating fields adjacent to the town, and built a storehouse for gunpowder and a jail.[35] Additionally, Mascaró reconnoitered the area of San Buenaventura in Nueva Vizcaya and drew plans for new buildings at that site. He checked the strength of the church of Charain and built a dam on the Yaqui River near the town of Onabas northeast of Guaymas. One of

---

[35] AGS, *Guerra Moderna*, 7241.

his more important projects was the building of a main irrigation canal at Pitic which put great tracts of land into cultivation.[36]

Mascaró's most ambitious project was one that never progressed any further than the planning stages. In 1780, he drew complete plans for the civic improvement of Arispe[37] that his 1781 report so desperately pleaded for. In his plan, Mascaró suggested urbanization projects that he would have liked to supervise. He included an elegant palace for the commandant general, an episcopal palace as well as cathedral, the *casa de moneda,* and a new city council building. All of these structures were to be arranged around or between four spacious squares and a large *alameda,* or public park. But the dream Mascaró blueprinted for Croix of an architectural phoenix at Arispe was too grandiose a scheme for available funds and the vicissitudes of the Commandancy.

More success came to another of Mascaró's civic designs. In 1783, Croix sent a letter to Gálvez explaining projected costs and itemizing funds collected for the royal treasury collection center newly established in the old mining town of Rosario.[38] Mascaró drew plans[39] for a sprawling single story building of neoclassic styling. It had two central

---

[36] AGI, *Audiencia de Guadalajara,* 517.

[37] "Plano General de la mision, y pueblo de Arispe, que S.M. en sus Rs. Instrucciones destina para capital de las Provs. Ints. de Nueva España," Arispe, Sept. 12, 1780, British Museum Mss. Room, Add. 17661b; and reproduced in Navarro García, *Don José de Gálvez,* following p. 400.

[38] Croix to Gálvez, Arispe, Dec. 1, 1783, AGI, *Audiencia de Guadalajara,* 518.

[39] "Plano de una Casa que se projecta construir para la Real Caxa nuevamente establecida en el Real de Rosario," Aug. 28, 1783, AGI, *Audiencia de Guadalajara,* 518; and reproduced in Navarro García, *Don José de Gálvez,* following p. 288.

patios behind the main entrance, ample living quarters, space for archives and offices, and doorman's quarters arranged within. The treasury sub-station building went up according to Mascarós specifications and activities headquartered there gave new impetus to the economy of Sonora.[40]

Besides his architectural plans, Mascaró also produced several strictly geographical maps. In 1778, Mascaró drew a handsome map of northern New Spain, based on Lafora's 1771 map.[41] Viceroy Bucareli had ordered it to help determine which areas of the Borderlands should be included in the new Commandancy General. The map was copied in 1779, possibly by Mascaró himself, with some additions to it.[42] Mascaró produced another similar map, but more elaborate, showing the routes of Anza from New Mexico to Sonora, and those of Croix, and Domínguez and Escalante, in 1782. This is the most detailed and refined of the three similar maps.[43]

One more map by Mascaró is known, but necessarily it was copied, or very largely based on someone else's work. It is of eastern Texas, which Mascaró never visited, thus excusing the many distortions and errors.[44]

---

[40] Navarro García, *Don José de Gálvez*, p. 544.

[41] "Carta o Mapa Geográfico de una gran parte del Reyno de N.E.," 1778, AGI, *Audiencia de Mexico* 346; and reproduced in Navarro García, *Don José de Gálvez*, following p. 336.

[42] "Mapa Geographico de una gran parte de la America Septentrional," 1779, British Mus. Mss. Room, Add. 17652b; and reproduced in García, *Don José de Gálvez*, following p. 336. *See also* illustration on page 157, herein.

[43] "Mapa Geografico de una gran parte de la America Septentrional," Arispe, July 29, 1782, British Mus. Mss. Room, Add. 17652a; and reproduced in Navarro García, *Don José de Gálvez*, following p. 336.

[44] "Mapa que comprehende la Provincia de Texas desde la mision de la

Croix was pleased with Mascaró's work in Arispe, even though many of his architectural designs were never realized. In 1782, Croix wrote to Gálvez in Spain praising the engineer. The commandant general was particularly impressed with Mascaró's dam construction, his maps, and erection of the gunpowder warehouse. For all these services, Croix heartily recommended Mascaró for promotion to captain. Gálvez passed the letter on to Corps officials without comment, and they notified Gálvez that in any case, Mascaró was about to be promoted because of vacancies on the list; and that there was no need to make a special case.[45] In due time, and strictly according to regulations of the rigid Corps, Mascaró was indeed promoted to captain and ordinary in 1783, receiving the good news in the spring of 1784.[46] At that time, Mascaró was earning 1,500 pesos annually, paid to him by the royal treasury through the Commandancy General.[47]

Less than a year after his promotion to captain, Mascaró received an order of transfer from the Interior Provinces to Mexico. Just a month afterwards, he departed Arispe, and left the Borderlands behind to continue his career in the capital. Mascaró

Espada hasta el Rio Misisipi," National Lby. (Paris), *Manuscrits mexicains*, no. 170, fol. 4; and reproduced in Navarro García, *Don José de Gálvez,* following p. 504.

45 Croix to Gálvez, Arispe, July 29, 1782; Gálvez to Abarca, El Pardo, Feb. 28, 1783; and Abarca to Gálvez, Madrid, Mar. 7, 1783, AGI, *Audiencia de Guadalajara,* 517.

46 [The king] to Croix, San Ildefonso, Sept. 28, 1783; and Croix to Gálvez, Arispe, Apr. 5, 1784, AGI, *Audiencia de Guadalajara,* 517.

47 "Estado que manifiesta los Liquedos Productos qe. en el año de 1784 han rendido a S.M. los Ramos de Rl. Hazda. establecidos en las Provas. de Sonora: y los Gastos qe. en ella sufre anualmente el Rl. Erario," Arispe, Sept. 9, 1785, AGI, *Audiencia de Guadalajara, 521.*

distinguished himself in Mexico as well. He worked on the pleasure park at Chapultepec, one of the governmental palaces, and the water system. He made repairs on the jail in the viceregal court. He also made a reconnaissance of the fortifications at Acapulco, and lent his talents to the Toluca road. His knowledge of his field was great and his drafting skills were considered outstanding by Miguel del Corral.[48] Mascaró steadily moved up the ranks during his service in Mexico City. By 1795, he was lieutenant colonel and second class engineer in command of engineers stationed on the Veracruz coast. With regular promotions for good service and longivity, Mascaró was brigadier and director sub-inspector of the Corps, second in rank in New Spain only to Costansó by 1809. It is in that year that Mascaró's name last appears in the rolls of the Corps of Engineers, and he probably died while still working in New Spain.[49]

The departure of Rocha and Mascaró from the Interior Provinces by 1785, did not mean the end of engineers' activities in the area. Croix, the viceroys, and other officials all realized the value and contributions of the skilled specialists in an area concerned primarily with defense, but also were grateful for their services in urban improvement and agriculture. Even before Mascaró and Rocha had left the

---

[48] AGS, *Guerra Moderna,* 7241; and AGN, *Provincias Internas,* 121.

[49] Marqués de Branciforte to Conde del Campo de Alange, Mexico, Nov. 31, 1795, AGS, *Guerra Moderna,* 7243; AGS, *Guerra Moderna,* 3794; and "Escalafón de 1809," *Memorial de Ingenieros del Ejército* (May 1908), pp. 347-51.

Interior Provinces, plans were being made to send another engineer to Sonora; not long after Rocha's dust settled on the road to the capital, a feisty Basque, Juan de Pagazaurtundúa, arrived in the north to fill his post.

6

Apache Warfare and the
End of the Colonial Period

Teodoro de Croix was appointed viceroy of Peru in 1783 in recognition of his excellent service in the Interior Provinces. A great change occurred in the pace of activities of the northern frontier of New Spain after Croix's departure. Croix's immediate successor as commandant general, California's Felipe de Neve, died a year after assuming his new office. Other factors minimized the accomplishments of the Commandancy after Croix left Arispe. An interim appointment to the commandant general's post allowed much of the commandant general's power to be delegated to the viceroy. Able Viceroy Bernardo de Gálvez died suddenly and was succeeded by less competent men while the Interior Provinces, split into two separate jurisdictions, lost much autonomy. Croix's farsighted action and planning were gone; divided interests and commands diffused the effectiveness of succeeding administrations.

Jacobo Ugarte y Loyola, former military governor of Sonora, became commandant general of the Western Interior Provinces in 1786. In his four years in office, he concentrated on the Apache problem to the exclusion of other activities that had occupied Croix and army engineers. Pedro de Nava followed Ugarte in office, and in 1793, the two divisions of the Interior Provinces were again united under one command. The authority of the viceroy was reduced and several provinces, including the Californias, were detached. Organization was finally stabilized until 1821, the end of the colonial period. But after 1783, engineers in the Interior Provinces were neither so active nor so necessary to administrative goals. Their participation was mainly directed toward helping in the fight with the Apaches. Their assignments were so diversified that their training and talents were dissipated through lack of overall direction. No longer were they called upon for major architectural and civic commissions, as Rocha and Mascaró had been; no longer did the engineers work so closely with the commandant general as they had with Croix; no longer were the engineers considered so important.

Decreased stature notwithstanding, engineers were still sent to the Interior Provinces and were still chosen with care for their assignments. In 1786, Manuel Pueyo was named to succeed Rocha in the Interior Provinces. He arrived in Mexico near the end of the year, just after his brother-in-law, Ruperto Luyando, an *oidor* of the Audiencia of Mexico, had died. Pueyo and his sister, Doña María, made special petition to the governing Audiencia asking that the engineer be allowed to accompany his widowed sister back to the family home in Zaragoza. In view of the special circumstances and the prestige of the deceased, the Audiencia made the emergency decision instead of going through regular channels. Pueyo was given temporary leave to escort his sister home at his own expense.[1] By the end of May, sister and brother had disembarked in Cádiz, and soon afterwards, the engineer requested permission to remain in Europe on grounds that he was about to move up to the class of ordinary, and only an extraordinary was needed in the Interior Provinces. By royal order, Pueyo was allowed to stay in Spain, and a subaltern was to be sent to his assignment.[2] In July of 1787, on suggestion from Corps officials in Madrid second lieutenant and assistant engineer, Juan Pagazaurtundúa, stationed in Veracruz, was named to the post in the Interior Provinces.[3] In October, word was received of Pagazaurtundúa's appointment by the viceroy, and extra money was granted to him for the trip north which he was to begin as soon as he recovered from an illness.[4]

Assigned to the Interior Provinces by default, Pagazaurtundúa was to spend a good deal of his time there trying to get transferred out. He had been born in Mexico, possibly the son of some official, since his service record lists his social standing as hidalgo.[5] In

[1] Audiencia de México to Marqués de Sonora, Mexico, Dec. 26, 1786, AGS, *Guerra Moderna*, 3805.

[2] Pueyo to Marqués de Sonora, Madrid, July 16, 1787; [the king] to Gerónimo Cavallero, San Ildefonso, July 22, 1787, AGS, *Guerra Moderna*, 3805.

[3] Antonio Valdés to Gerónimo Cavallero, San Ildefonso, July 22, 1787, AGS, *Guerra Moderna*, 3070.

[4] Manuel Antonio Flores to Antonio Valdés, Mexico, Oct. 22, 1787, AGS, *Guerra Moderna*, 3805; and Oct. 23, 1787, AGN, *Virreyes*, 142.

[5] Service record, Jan. 1, 1787, AGS, *Guerra Moderna*, 5837.

Spain in 1774, at the age of nineteen, he joined the Infantry Regiment of Soria and advanced to second lieutenant. He proved his valor during the blockade in Gibraltar from 1779 to 1782, and studied mathematics at the Academy of Barcelona before returning to Mexico with his regiment. On May 6, 1785, after twelve years of army experience, Pagazaurtundúa was named to the Corps. He served in Mexico and Veracruz before being assigned to the Interior Provinces.[6] No more than a year after Rocha had departed, Pagazaurtundúa arrived in the Borderlands to replace him. Very soon afterwards, he was promoted to lieutenant and extraordinary engineer.[7]

In the next five years, Pagazaurtundúa was occupied with various commissions to maintain and repair ecclesiastical and government buildings. He worked on the parish church at Chihuahua and the church at the mining center of Santa Eulalia. He drew at least one map, reconnoitered the presidio sites in Nueva Vizcaya, and spent time in New Mexico.[8] During this period, Pagazaurtundúa was unhappy with his assignment. He objected to the rigorous work, to high frontier prices, and to the climate, which he maintained had ruined his health. In fact, the engineer stated in a 1793 letter to the viceroy that though still a young man, he had aged in the Interior Provinces so that he looked many years older.[9] This is one of many characteristic exaggerations Pagazaurtundúa

[6] AGS, *Guerra Moderna,* 3794.

[7] Gerónimo Cavallero to Antonio Valdés, the Palace, Dec. 23, 1788, AGS, *Guerra Moderna,* 3806.

[8] Pagazaurtundúa to Conde de Revilla Gigedo, Guadalajara, Apr. 19, 1793, AGN, *Provincias Internas,* 121; AGS, *Guerra Moderna,* 5837.

[9] AGS, *Guerra Moderna,* 7240.

made in his pleas for transfer and promotion. He was 38 at the time he wrote the letter, hardly a young man by actuarial standards of the day. At any rate, Pagazaurtundúa requested reassignment, since he had been at his post the mandatory five years. At the time, he was in Guadalajara, temporarily replacing engineer Narciso Codina, who had been transferred.

While in Guadalajara, Pagazaurtundúa continued Codina's work on a church and the water supply system for the city. He was allowed to remain there instead of returning to the north, and Miguel del Corral confirmed his bad health. Additionally, 25 pesos monthly was to be paid to him in compensation for municipal works, according to Corps regulations.[10] Six months after Pagazaurtundúa had left the Interior Provinces, a replacement for him was proposed, and Pagazaurtundúa was told to return to Spain as soon as José Cortés arrived to relieve him.[11]

At the beginning of 1794, Pagazaurtundúa was still in Guadalajara, and Corral reported that his health was good.[12] He left New Spain in 1796,[13] and was quickly reassigned to Mexico; but he did not go immediately. During his first months in Spain, Pagazaurtundúa wrote a description of the Interior Provinces for Lieutenant General Luis Huet of the army of Andalucía, former director of the Corps of Engineers. The description[14] is short, and suffers

---

[10] Revilla Gigedo to Conde del Campo de Alange, Mexico, May 29, 1793; Sabatini to Campo de Alange, Madrid, Oct. 3, 1793; Sabatini to Campo de Alange, Madrid, Nov. 16, 1793, AGS, *Guerra Moderna,* 7240.

[11] AGN, *Provincias Internas,* 121.

[12] AGS, *Guerra Moderna,* 7241.

[13] AGS, *Guerra Moderna,* 7243.

[14] "Succinta Descripcion de las provincias internas," Cádiz, Mar. 28, 1797, BCM, ms. 5-2-1-7. See Appendix H.

from its brevity. Pagazaurtundúa wrote in broad generalizations which conveyed very little information about the state of defense, the economy, or the people in the Interior Provinces. He listed the main activities in the area as agriculture, mining, and Indian fighting, but elaborated little, and made no recommendations whatsoever for improvement. Pagazaurtundúa thought that the presidial troops were doing an excellent job in controlling the Apaches. Without them, he noted, the sparse Hispanic population there would have disappeared long ago.

The purpose for which this description was written is not known with certainty; Huet had ordered it, perhaps calling Pagazaurtundúa to account for his time spent in the Interior Provinces. At the end of the report, Pagazaurtundúa explained that he had worked on several maps of all of New Spain for Viceroys Flores and Revilla Gigedo, as well as for the Commandancy. But, he said, since he was the only engineer in all of the northern region and since he had many other things to do, he had not had time to draft a map to accompany this report, which would have supplemented what he had put in writing. However, he did send along a copy of the military footing in Sonora, Nueva Vizcaya, and New Mexico.

Only one map of those made by Pagazaurtundúa has been located.[15] Dedicated to Manuel Godoy, it was made in America in 1803. It is large, attractive,

---

[15] "Nuevo Mapa Geográfico de las Provincias de la Sonora y Nueva Vizcaya de la America Septentrional," 1803, British Mus. Mss. Room, Add. 17660b; and reproduced in Navarro García, *Don José de Gálvez,* following p. 456. *See also* illustrations on pages 175-77, herein.

neat, and very detailed, but contains notable errors: Sonora and Nueva Vizcaya appear to be a thickly wooded mountain region, but the river courses are acurately depicted; Durango is mislocated; jurisdictional areas are confused. Obviously, the map was not based on first-hand knowledge or, if it was, time had dulled Pagazaurtundúa's memory of the Interior Provinces.

But before the 1803 map was produced, at about the time Pagazaurtundúa wrote his description, he had been reassigned to Mexico. Before he could leave Spain he was taken prisoner of war by the British and was held four months before being released at Gibraltar. From there, the engineer walked to Cádiz, and in that port, petitioned the king for reimbursement of funds he had spent while a prisoner. The king denied Pagazaurtundúa's petition, but granted funds for the engineer's return trip to New Spain as soon as it might be possible.[16] Meanwhile, the viceroy in Mexico had been informed that Pagazaurtundúa, assigned to his command, had been taken prisoner and that he was in Cádiz on his word of honor until hostilities should cease. Viceroy Miguel de Azanza requested a replacement, and extraordinary engineer José Aloy was duly transferred to Mexico from his post in Caracas.[17]

In the fall of 1802, Pagazaurtundúa finally arrived in New Spain, having remained three years as a

---

[16] Pagazaurtundúa to the king, Cádiz, Apr. 16, 1799; [the king] to José de Urrutia, Aranjuéz, May 26, 1799, AGS, *Guerra Moderna,* 7246.

[17] Miguel de Azanza to Juan Manuel Alvarez, Mexico, Sept. 26, 1799; [the king] to Azanza, San Lorenzo, Oct. 22, 1800, AGS, *Guerra Moderna,* 7247; and BCM, vol. 57.

prisoner of war.[18] Before being taken prisoner by the British, Pagazaurtundúa had petitioned the king for command of the Guaraní province in the Viceroyalty of Río de la Plata, but the request was denied.[19] After his years as prisoner, at least one attempt to better his position with a civil post, and complaints about his assignment in the Interior Provinces, Pagazaurtundúa was back in New Spain. He accepted his assignment with good grace at this point and remained in Mexico for the rest of his career. He had been promoted to the new rank of brigade sergeant major by the time he arrived in Mexico. By 1805, he had advanced to the rank of lieutenant colonel and after 1818, his name disappears from the records.[20] He probably died while serving in New Spain.

From the nature of available documents, Pagazaurtundúa appears to have been a more interesting personality than some of the other engineers who served in the Borderlands. But at the same time, his participation in engineering activities on the northern frontier was neither so large nor so important as that of Lafora, Costansó, Mascaró, and others. Similarly, the last two engineers concerned with the Borderlands during the Spanish period contributed less than earlier pioneers.

José María Cortés de Olarte was named to replace Pagazaurtundúa, when he received his long sought-

---

18 Pagazaurtundúa's name appears on the Corps list of officers in the Indies in 1803 showing his embarkation date, after absence from the rolls for the three years preceding, AGS, *Guerra Moderna, 3794.*

19 Pagazaurtundúa to the king, Cádiz, Mar. 27, 1798; [the king] to Pagazaurtundúa, Aranjuéz, Apr. 4, 1798, AGS, *Guerra Moderna, 7245.*

20 AGS, *Guerra Moderna,* 3794; *Calendario Manual y Guia de Forasteros,* pp. 127-28.

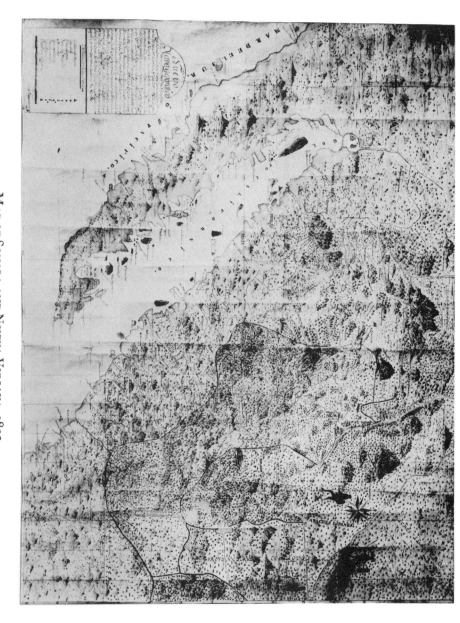

Map of Sonora and Nueva Vizcaya, 1803

Prepared by Juan de Pagazuartundúa, the map is well drawn but with many errors.

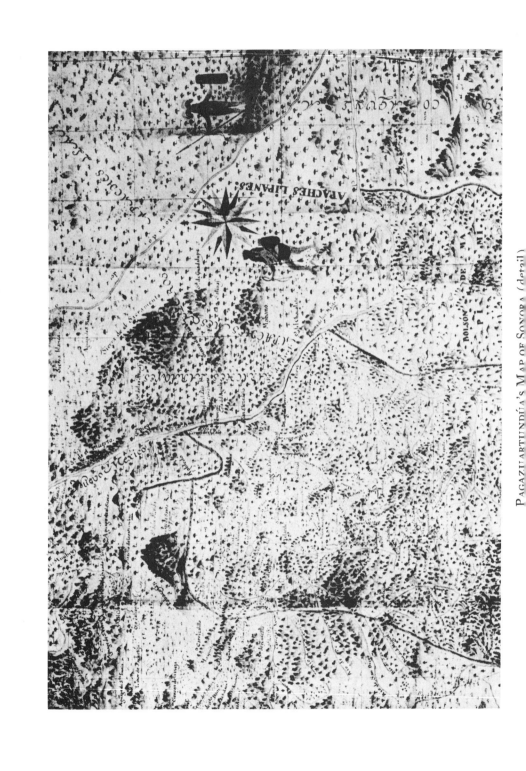

PAGAZUARTUNDÍA'S MAP OF SONORA (detail)

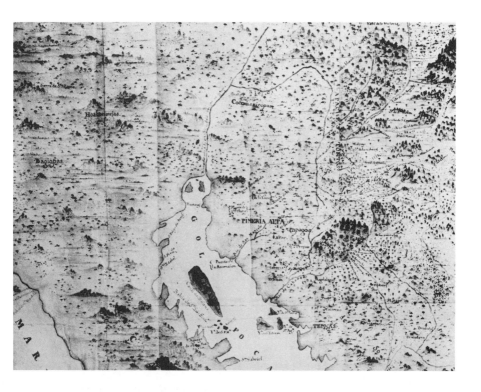

Pagazuartundúa's Map of Sonora (details)

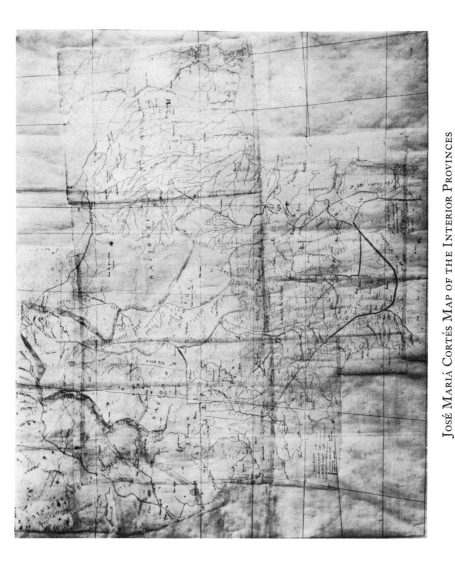

Josñ Mariá Cortés Map of the Interior Provinces
A rough sketch map having many innacuracies.
See page 184. Courtesy, Bancroft Library.

after transfer out of the Interior Provinces.[21] At the time, Cortés was 24 years old, having joined the Regiment of Toledo nine years before. While a cadet, he attended the mathematics Academy at Céuta across the strait from his home in Tarifa. In 1788, Cortés was authorized to go to Madrid for entrance examinations for the Corps of Engineers, and he was admitted as second lieutenant on January 23 of the following year. He served in Galicia as assistant engineer before being assigned to Cádiz where he received the order transferring him to New Spain.[22] With passports arranged for the engineer and one servant in Cádiz,[23] Cortés left Spain in May of 1795, after having been promoted to lieutenant and extraordinary engineer.[24]

Shortly after his arrival in the Interior Provinces, Cortés solicited funds from the king to defray expenses of his long march from Veracruz to Chihuahua. Commandant General Pedro de Nava seconded the engineer's request, stating that the king had ruled favorably on a similar petition made by an engineer some years previously. The king responded with an outright grant of 400 pesos and declared the matter settled, even though Cortés' expenses might have been greater than that sum.[25] The ample size of the stipend

---

[21] [The king] to Commandant General of the Interior Provinces, San Lorenzo, Dec. 10, 1793, AGS, *Guerra Moderna,* 7240.

[22] Service record, 1796, AGS, *Guerra Moderna,* 3794; Juan Cavallero to Gerónimo Cavallero, Madrid, Oct. 21, and G. Cavallero to J. Cavallero, Madrid, Oct. 24, 1788, AGS, *Guerra Moderna,* 3029.

[23] Sabatini to Campo de Alange, Madrid, Nov. 16, 1793, AGS, *Guerra Moderna,* 7240.

[24] BCM, vol. 57; and AGS, *Guerra Moderna,* 3794.

[25] Pedro de Nava to Campo de Alange, Chihuahua, May 3, 1796, AGS, *Guerra Moderna,* 7243.

probably satisfied Cortés and gave him more incentive to continue his service to a generous king.

Little is known of Cortés' activities in the Interior Provinces except what can be gleaned from a lengthy essay on the area he wrote in 1799.[26] The document has neither the air nor the form of an official report of the Corps of Engineers. It is subjective in that Cortés limited himself to his observations, his reflections, and only the most trustworthy secondary sources. He wrote his *Memorias* after having been in the north for three or four years, but with a more distant perspective. The document's importance, though unofficial, lies in Cortés' impressions of the Interior Provinces, and particularly his view of Apache affairs, a subject on which he remained one of the king's enlightened experts.

The first section following Cortés' preface is a typical description of the area. It is especially complimentary concerning the areas of Nueva Vizcaya and Sonora. As Mascaró had done in his description of the Arispe area, Cortés noted that the soil was rich, and that all sorts of woods for construction and furniture were available in the mountains. Even more valuable was the wealth that could be extracted from the provinces in silver and gold. Cortés thought it unfortunate that only a few rich *hacendados* controlled the resources of the huge Interior Provinces, while the rest of the people, poor as they were, seemed content with what little they had. Cortés estimated the population of the Interior Provinces

26 "Memorias sobre las Provincias del Norte de Nueva España," 1799, ms. copy of original in the British Mus., Univ. of Ariz. microfilm no. 61.

at 300,000, a low figure compared to Alexander von Humboldt's reckoning of a few years later.[27]

In discussing military affairs in the northern regions, Cortés stated that there were 3,099 soldiers encompassing twenty presidial companies, five mobile companies, and three companies of Pima and Opata auxiliaries. He attributed to the frontier soldiers phenomenal capabilities and endurance, insisting that they could be out in the field on campaign for fifty days without shelter and would neither tire nor become ill. He said they were constant, long-suffering, humble, and obedient, and had the singular virtue of acting courageously in battle even when their officers did not set a good example. Despite their loyal and invaluable service, Cortés bemoaned the fact that the presidials were furnished with poor rations and were subjected to useless practice maneuvers and drills. Somehow, Cortés observed incredulously, the soldiers were able to maintain their energy and *espirit de corps*. His only criticism of the soldiers was their marksmanship. With more effective fire power, he concluded, bloodshed all along the Apache frontier could be lessened. Although the Indian arrow was fast, Cortés correctly noted that its range was short, and that of the musket much greater. Had the Spaniards possessed greater accuracy in their fire power, the Indians would have been afraid to fight against such superiority and hostilities would decrease.

---

[27] Humboldt said the Interior Provinces had a population of 559,200 in 1803. Juan A. Ortega y Medina, ed., *Ensayo Político sobre el reino de la Nueva España*, p. 357.

The *establecimientos de paz,* the new system of settling Apache bands and attempting to civilize them by making them dependent on the Spaniards for liquor, among other items, notably impressed Cortés. He considered absurd the prevalent idea that Apaches were not capable of living a peaceful and civilized existence. "They love peace and fear losing it," he said. But since 1786, when warfare on them had been intensified, many *rancherías* truly desiring peace had fallen. Cortés admitted that many Apaches had fled to mountain retreats, but he defended their flight as justified. Although they had committed crimes, he explained, the Spaniards were as guilty of wronging the Apaches as the Indians were guilty of hurting the Europeans.

Cortés agreed with the commandant general and other officials that the Apache had to be pacified, but he suggested a method other than relentless warfare. He favored a policy of peace through purchase, which had been used successfully with the Chichimecas two centuries before and was to be employed again by the United States government in the next century. Since hunting and raiding were the Apaches' only means of support, Cortés thought they should be given the necessities of life and settled near Spanish communities. Cortés rhetorically supported his plan by proclaiming that the Spanish blood the Apaches could spill in one week on the raiding warpath was not worth any amount of money it might take to keep them supplied with food and other goods.

To back up his ideas Cortés cited the success already achieved with the *establecimientos de paz.* He wrote that Apaches were helping supply food and clothing for their families. They volunteered to help the soldiers as scouts, they hunted game, and they guarded their identification papers with care and pride. When volunteers were needed to serve as auxiliaries, every Apache wanted to be chosen to participate. On campaign, they served as guides, hid the troops in safe, undetectable places, spied, and then practically directed the attack. In battle, Cortés reported, the Apaches fought like lions even against their own kinsmen, and when overpowered by the enemy, they died bravely in the service of Spain, true and loyal mercenaries.

Cortés said there were many documents that supported his stand. He thought that attracting and settling the Apaches should have been the most important objective of the government of the Interior Provinces. It was in the interest of religion, wise government, and humanitarianism to take up this project. Success would bring stability to the frontier, and possession of the area would finally be secure, permanent and uncontested. Further, if the Indians were treated not only with justice, humanity, and trust, but also with severity when they did wrong, after twelve or fifteen years they would be happy with their new sedentary way of life. "The grandchildren of the savage Indians, so ferocious, destructive, and bloodthirsty," Cortés prophesied, would be "useful vassals, humble and pious, just like other peoples."

Cortés long recommendation for purchasing Apache allegiance was aimed at increasing the Hispanic population of the frontier, the same goal which Costansó and others had hoped to accomplish for defensive reasons. Cortés said that with the Apaches controlled, colonists would be encouraged to settle where they could pursue mining, farming, and commerce with less risk. Basically, said Cortés, the Indians were humble, and loved the king! But most frontier leaders were incapable of enlightened command and continually had frustrated the peaceful desires of the Apache. This criticism of the military leadership on the frontier seems to have been Cortés' main point, and possibly the rationale behind his effusive praise for presidial and Indian alike. What Cortés hoped to gain is almost unfathomable. His liberal thinking as regards the nature of the Apache was most unusual, unorthodox, and unprecedented, as well as somewhat naive.

Possibly drawn to accompany his 1799 *Memorias,* was a map of the Interior Provinces made by Cortés in the same year.[28] The map is rough and incomplete, really a mere sketch compared to others of the period. It shows the major geographical features, but with very little accuracy. Indian tribe locations and presidios are marked, but rivers flow in greater profusion than that with which nature graced the Interior Provinces. The map appears to be a rough draft, or a preliminary version for a later effort, but no improvement on this basic sketch has been located.

After writing his *Memorias* and drawing his map in 1799, Cortés remained in New Spain for several years. While still in the Interior Provinces, he was

---

28 "Mapa Geogo. de la Ote. de la America Sepl. aumentado y corregido por dn. José Cortés Yngo. de los Rs. exercitos," 1799, British Mus. Mss. Room, Add. 1765c; and reproduced in Navarro García, *Don José de Gálvez,* following p. 504. *See also* illustration on page 178, herein.

promoted to captain under the new organization of
the Corps in 1802. He left New Spain in 1803, and
was promoted twice in close succession shortly after
his return to the peninsula. In recognition of merit
for his services on the frontier, he moved up first to
sergeant major and then to lieutenant colonel in the
Corps by 1805.[29] Nothing more is known of Cortés'
professional activities, but an interesting bit of personal information has been uncovered. On January 2,
1809, Cortés was arrested in Sevilla, apparently because he was thought to be away from his post without
proper papers. The engineer protested the injustice
of his arrest while he was being escorted to Granada
to appear before the provincial military chief. He
maintained that he had been given a thirty-day leave
although he did not have orders with him.[30] Cortés
must have been cleared of charges against him because less than six months later he was granted license
in Sevilla to marry Doña Catalina Ximénez Briton
y Landa. At the time, 42 year-old Cortés was in Cádiz
at his post.[31] Possibly his visit to Sevilla a few months
earlier had been one of romantic purpose to plan a
summer wedding. No other mention of Cortés has
been located in the archives. Nor have documents
been located showing that he was replaced in the
Interior Provinces.

Though he never set foot in the northern region of
New Spain, one other royal engineer vicariously par-

[29] BCM, vol. 57; and AGS, *Guerra Moderna,* 3794.

[30] Antonio Cornel to Gregorio de la Cuesta, Sevilla, Jan. 7; Cornel to
Martín de Garay, Jan. 8; and Cornel to Cuesta, Jan. 10, 1809, AHN, *Estado,*
45.

[31] AGM, *Expediente personal de José María Cortés.*

ticipated in the area in the last days of Spanish control. Juan Camargo Cavallero, born in the south of Spain in 1759, had served in the Cantabrian Infantry Regiment for fifteen years before entering the Corps of Engineers in 1787. He advanced rapidly, serving in all the coastal areas of Spain, and arrived in New Spain in 1791 as official replacement for Narciso Codina, the engineer Pagazaurtundúa temporarily replaced in Guadalajara. Camargo carried out important commissions in Tlaxcala and other areas of the interior, but was finally firmly stationed at Veracruz where he was in charge of construction and repair of the harbor fortification, San Juan de Ulúa.[32]

By 1815, Camargo was full colonel and one of the highest ranking corpsmen in the Americas. Still stationed at Veracruz, he was ordered by superiors in Spain to write a report on the state of New Spain. Already besieged by colonial rebellion for five years, Spanish officialdom was justifiably concerned about its richest and most important possession, New Spain. The resultant report by Camargo[33] is an extremely complete if overly optimistic document. In over a hundred pages of manuscript, Camargo started with a physical description, province by province, in the customary manner of such reports. Then he included a brief history of the realm beginning with the Cortesian conquest three hundred years before and described the many Indian tribes and racial castes

---

[32] Service record, Dec. 31, 1795, AGS, *Guerra Moderna,* 7241; Passport of Camargo, his wife, two children, and one servant, San Lorenzo, Oct. 4, 1790, AGS, *Guerra Moderna,* 7237.

[33] "Memoria sobre el Reyno de Nueva España Provincias Internas y Californias," Veracruz, Oct. 24, 1815, BCM, ms. 5-2-4-3.

that made up the population of the realm. He gave complete financial and economic reports for the entire kingdom for the three years immediately preceding outbreak of the war for independence. Additionally, this experienced engineer reported on the pre-war state of military defense, dealing with external attack, preparedness, and internal strength. He went on to report on the nature of the insurgents and classified them in groups. His tone generally conveyed the idea that with a modicum of wise action Spain really did not need to worry about the success of any rebellion.

Camargo's report did not add anything new to information already compiled years earlier on the western Borderlands. He was so preoccupied with the war that he did not even mention the Apache Indians in the Interior Provinces, the subject which had been of primary concern in that area for the last half-century or more. Very briefly, he reviewed the presidial situation along the northern frontier, but really only in passing since Spain at the time could well have afforded to lose the desolate northern half of New Spain if only central Mexico could be retained.

The very brief mention of defenses in the Borderlands by Camargo is significant because it is the last known documentary reference to this area by a member of the Royal Corps of Engineers. The Corps, like the rest of the Spanish army stationed in New Spain, was no longer permitted to invest time and talent in the far north. The loss of an empire was at stake; corpsmen had been sent to the Interior Provinces with regularity and with high regard for their importance during the reign of Charles III, their participation in the northern provinces dwindled under Charles IV, and finally disappeared with the cataclysm of the war for independence.

All over New Spain and, indeed, all over Spanish America, the decline in administrative efficiency had begun perhaps about 1790 and had accelerated rapidly after the turn of the century. With war in 1810 and Spain torn asunder by events related to the Napoleonic invasion, colonial New Spain faced havoc, confusion, and imminent disintegration. Spain lost its longtime firm grip on America, and the activities of the Corps of Engineers, or lack of them, in the far north, reflected these serious preoccupations.

The accomplishments of the Corps of Engineers in the Spanish Borderlands are in the past and all but forgotten. It is not possible to identify or quantify with assurance the Corps' contribution to the administration of the vast northern frontier. But it is possible to weigh and measure, or at least consider, this hitherto ignored institution of the frontier. Education, specialized training, and Corps indoctrination assured the engineers of being enlightened, aware men of their times. Technical expertise was certified by rigorous examination for entrance to the Corps and constant evaluation thereafter. Military training in the construction, location, and defense of fortifications assisted the presidial commanders and policy-making officials responsible for the "DEW" line of Spain's northern frontier. Battle experience provided engineers first-hand acquaintance with combat warfare and prepared them to better understand the resident enemy of the western Borderlands. Peninsular origin and cultural orientation of the engineers resulted in current European ideas and opinions being transported to isolated regions that suffered from infrequent and irregular contact with the mother country. Documentary material – the reports, diaries, recommendations, and maps that the engineers produced – contributed geographical, social, economic, demographic, and above all, military knowledge about the frontier regions. At the very least, then, the Royal Corps of Engineers helped to create, maintain, enhance, and defend Spain's western Borderlands.

Appendices

Appendix A

Ranks and Classes of the Corps of Engineers, 1711-1768

Class	Rank	Quota
Ingeniero General Engineer General [1]	[Unspecified]	[1]
Ingeniero Director	Brigadier, Mariscal de Campo, Teniente General [2]	10 [3]
Director Engineer	Brigadier, Field Marshal, Lieutenant General	
Ingeniero en Jefe Chief Engineer	Coronel Colonel	10
Ingeniero en Segundo Second Class Engineer	Teniente Coronel Lieutenant Colonel	20
Ingeniero Ordinario Ordinary Engineer	Capitán Captain	30
Ingeniero Extraordinario [4] Extraordinary Engineer	Teniente Lieutenant	40
Ingeniero Delineador Draftsman Engineer Ayudante de Ingenieros [5] Assistant Engineer	Alférez Ensign Subteniente Second Lieutenant	40

---

[1] English translations as utilized in the text.

[2] Ranks corresponding to classes were set by royal decree of 1756.

[3] Class quotas were established by the Ordinance of 1768.

[4] This class was added by royal decree in 1724.

[5] Nomenclature for this class and rank was changed by the Ordinance of 1768.

Appendix B

Description of the Provinces of
Culiacan, Sinaloa, and Sonora [1] [Fersen, 1770]

The territory known until now by the name of the Government of Sonora and Sinaloa, extends from the southeast to the northwest between the western range of the Sierra Madre and the corresponding seacoast. It commences at the Acaponeta River, which divides it from Nueva Galicia, and ends at the Presidio of Tubac and the Mission of San Xavier del Bac bordering on the territory of the gentile Sumas, Tepocas, and other neighboring nations on the Gila River. On the northeast it borders on the realm of Nueva Galicia, and on the southwest, with the Red Sea of Cortés, or the Gulf of California. Its vast expanse of more than 300 leagues [2] in length, and from eighty to 120 in width, includes diverse types of terrain, with distinct climatic conditions; but it may be taken as a general rule that all the coastal regions are much hotter than Mexico, those in the Sierras very cold, and those in between quite mild and temperate. Many perennial and abundant rivers irrigate all of that country and make it generally fertile for cattle, grass, fruits, and everything necessary for natural life, especially on the fertile plains and in the river valleys, which is where the largest populations are usually found. Away from the rivers the terrain tends to be barren, and much more so near the beaches.

There is no textile manufacture in these provinces, either of wool or cotton. Artisans are scarce even for the simplest and most essential crafts, and everything needed for a comfortable, civilized existence is transported from Mexico and Guadalajara by mule, at great cost and danger.

All the territory to the Yaqui River is pacified, or reduced, and has been secure for many years so that there is nothing there to fear

---

[1] The Province of Ostimuri is also described and treated as a district separate from Sonora.

[2] The measurement of a common league varies from 2.3 to nearly 3 miles.

except perhaps an occasional small rebellion among the tame Indians, which good government curbs and for which Spanish settlements serve as a check. But from there the country is somewhat
exposed to harassment from the gentiles, and it is much more
dangerous in the far reaches of Sonora, not so much from the native
tribes as from the wandering Apaches.

The chief of the territory is a governor to whom the captains of
the presidios in military affairs, and the lieutenants and the high
court judges of the districts in political affairs, are subservient. For
ecclesiastical matters the area belongs to the Diocese of Durango or
Guadiana and the sacraments are administered by secular priests in
the reduced areas and by regular missionaries in areas that are not
yet completely converted. The government of this territory was
divided into four main provinces, the first being Culiacán, to which
the small province of Maloya is attached. Its boundaries are the
Acaponeta River on the southeast, the Mocorito River on the northwest, and on the northeast and southwest, the Sierra Madre and
the coast of the Gulf of California are common boundaries with the
whole territory. Its capital is the old Villa of Culiacán, a place of
medium population located on the bank of the river that bears its
name, which is quite abundant and is situated almost equally distant
from the sea and the Sierra. The province also has the small Villas
of San Sebastián, San Xavier, San Ignacio, and some other Spanish
settlements, and many more of Indians. It also has cultivated crops,
and cattle and livestock ranches. There are few sheep and little
wheat, but many other grains and fruits are produced, and there
are beautiful woods. Its proximity to the sea makes possible a prosperous business enterprise in salt.

The following mining districts are also located there: Rosario,
Pánuco, Copala, Plomozas, Cozala, Palo Blanco, Guadalupe, and
El Cajon; and many others along the boundaries, but smaller and
less important. The District of Rosario is the most important after
Guadalajara in all this interior territory. Its early boom attracted
a large population, which because of its favorable location and its
enterprise, has kept it going despite the deterioration of its mines.
It is 240 ordinary leagues distant from Mexico, and 115 leagues on
a good road from Guadalajara, and eighty from Durango, crossing
the Sierra over rocky and precipitous trails. It is located on the

banks of a river of the same name at the foot of the Sierra Madre, and it is fifteen leagues from the sea; therefore it is directly on the Camino Real of Sonora. The town is well arranged with four straight streets of good houses, about 600 *varas* [3] long, and five cross streets about 250 *varas* long, apart from the outskirts that surround it. There are about forty large drygoods stores and a greater number of grocery stores and from four to five thousand inhabitants could be supported. The merchants of Rosario supply all the miners of the province of Culiacán and give them credit without gold or silver. So although this district does not produce on its own more than 20,000 marks of silver and perhaps a thousand ounces of gold annually, because of its trade it attracts almost all of the silver and gold of this and the other interior provinces.

Fifteen leagues to the northeast of Rosario are the mining districts of Pánuco and Copala which produce 40,000 marks of silver annually thanks to the skilled and constant labor of the mines under Don Xavier de Viscarra, Marqués de Pánuco, a native of Rosario, where he has just built a magnificent church. The rest of the mining districts listed above are much less important, but are still producing plenty of silver and gold, and in all of them mining operations differ very little from the area around Mexico City, nearly all the silver being extracted by means of mercury.

On the Mocorito River, the Province of Sinaloa begins, and it ends on the Mayo River. It is flatter, larger, and has a more healthful climate than the preceding one, but it is not so well settled and there is much less traffic and less territory. Grains are plentiful, especially corn, which is of the best quality and cheaper than in any other of these regions, so that the yield is fifty per cent greater than in the Province of Culiacán and Nueva Galicia. Also, this area abounds in cattle and horses and other kinds of livestock. Its capital is the Villa of Sinaloa, which has a medium-sized population, and is laid out very poorly. The people are very poorly dressed, and unemployed, which is one of the reasons why it was abandoned as the seat of government, which it had been previously. Twenty-five leagues to the northwest of Sinaloa is the Villa of Cadereita, or Montes Claros, still better known as the Villa del Fuerte for the rushing river of the same name. This is a place with a considerable

---

[3] One *vara castellana* is equal to 32.09579 inches, or 0.835905 meters.

Spanish settlement, and there are also several pueblos like Bair-aguato, and many other places where there are Indian ranches and rural dwellings of miserable people.

The most important mining district of this province is the one at Los Álamos, which is about fifty ordinary leagues from Sinaloa, ninety from Culiacán, and about 160 from Rosario. It is an important place, reasonably well populated although the houses are widely scattered and its commerce consists of six or eight mercantile establishments. It is located in a low spot at the foot of very high mountains, and therefore has a very unhealthy climate. A contributing factor is the lack of drinking water in the dry seasons when it has to be drawn from wells. In the same neighborhood are the mining districts of Aduana, La Quintera, Cerro Colorado, and other small ones, but 28 leagues to the southeast of Los Álamos is the District of Siribipa, of medium population and very steady in the production of silver ores of high grade.

The gold mines of Tanaran, El Satac, and Bacaiopa also belong to this province, as do the Zapte silver and gold mines, San Juan Nepomuzeno, and Arizona where a few years ago a rock of pure silver was uncovered and there was a question as to its value. Also, there are the mines of San José de las Cruces, Tenoriba, Monzerrate, Los Tajos, Agua Caliente, Los Molinos, Los Cercanos, and Suristado. The majority of these mines are deserted, not presently being worked, not because there are no veins of good silver, but because they were badly worked from the start. They are either impossibly blocked by the debris from their workings or by floods, and would have to be worked by deep shaft or explosion, or other means too expensive to be undertaken by these poor people. It should be understood that from this province northward, the fine art of working the mines and refining metals is largely unknown. The placers of Bacubirito, ten leagues to the southeast of Sinaloa, have become famous these last few years for the very considerable quantity of gold they have produced. There are also other placers among the gold mining districts near the Fuerte River such as Bamicori, Lecorato, Agua Caliente, and others.

The province of Ostimuri is the terrain between the Mayo and Yaqui Rivers, which makes it even more fertile than the area previously described, and in all the rest much the same. Its population consists of Indian pueblos on the banks of the rivers and mining districts in the Sierra. The Mining District of Ostimuri, which gave the province its name, is the least important one today, and Trinidad is now most important. Also of considerable importance in this province are the districts of Rio Chico and Baiorcca, both well populated and in operation. The rest of the districts in the province are Concepción, Guadalupe, San Antonio, Candelaria del Sobia, El Potrero, Tacupeto, Guimanazorra, and Malpillas, and the gold placers at Palo Blanco, Potrenillo, and others. These have little or no work going on for the reasons explained above, but their metals are of high quality especially in the mines near Tauramana, a little known, rough part of the Sierra, as yet unsettled despite its long reputation for riches.

On the Yaqui River begins the province especially associated with the name of Sonora. It is undoubtedly the largest, the most fertile, and the richest of all the provinces that comprise this governmental jurisdiction. Sonora has a very healthful climate, especially away from both the sea and the Sierra. Its produce is very good, especially wheat, wool, and horses, which are thought to be the best in our America. This is accomplished on the banks of the perennial rivers and the valleys they water, because at some distance from them, near the sea, there are some barren stretches, where water is scarce.

Sonora is divided by a number of valleys which form the lowest part of the foothills of the Sierra and bear the names of the Spanish settlements there, such as the Sonora Valley, the Santa Anna, the Tapache, and the Ruipa, where there is also a considerable number of Indian settlements which were missionized only very recently. The governor resides at the Presidio of San Miguel de Horcasitas. Besides the latter, Sonora contains within its boundaries the Presidios of San Carlos de Buenavista on the Yaqui River, Altar, Tubac, Terrenate, and Fronteras, earlier known by the name of Corodeguachi.

The mining district of San Antonio de la Huerta, or Las Arenas, is the most notable place in Sonora because of its commerce. It is located on the west bank of the Yaqui, and is well laid out, with a closely settled population and it is rationally organized. It supports about 25 drygoods stores and other businesses. Each year 400 to 500 mules arrive laden with goods from Europe, Mexico, Puebla, and Guadalajara. Its location, which is in the most fertile and settled, and least exposed place in the province, boasts all the advantages, and is consequently the place where the miners gather to barter their silver and gold for the effects they need, and where the country people come to sell their grains and other products. The other mining districts currently operating in Sonora are La Ventana, San Miguel, San Francisco, El Carrizal, Nacatabori, Tonibari, Bacuachi, Tacombavi, Nacarri, Chinapa, Basochupa, Bacanuchi, Babicaneza, Motepore, Matape, Opodepe, Nacameri, San José de Gracia, Sanacachi, El Aguaje, and Aigano. Distances and locations are marked on the map which accompanies this description. The gold placers are almost as common as silver veins, and the one at La Cieneguilla at the boundary of Sonora could be used as an example of all the others. But the majority of these mines are not being worked at present and those that are worked are not producing as much as they should. The main reason is lack of development and settlers, and in some cases, the justifiable fear of harassment by the Apaches.

This condensed description is based on original memoranda, the most recent information from the provinces, and data I obtained for myself in the neighborhood from experienced and educated people. Obviously, it shows that these provinces are well endowed by nature to promote industry and labor in happy combination, and their potential prosperity is also undeniable. But more often than not, the attainment of this prosperity is poorly understood, since its source is under the ground, and therefore requires application of all the means available for this kind of development, whether applied to a greater or lesser degree.

Looking carefully at the western stretch of mountains called the Sierra Madre, a range of the great Cordillera (the world's largest) that stretches the length of both Americas, one can see that its whole extent of 300 leagues from Bolaños to the Presidio of Fronteras is dotted with mineral deposits that have been discovered less often by search than by accident. Only a few men of little intelligence have seen them, undeveloped as they are in the midst of a thousand obstacles and perils. What would happen if they could be treated like the mines in this area? Assuming this is the objective of the wise intentions of the government, one of the measures that can now be applied is the establishment of new revenue collection stations closer to the mining districts and consequently more convenient to the miners, for the purpose of showing their metals and obtaining mercury, and at the same time to prevent losses. The place best suited for this purpose appears to be the Real of Rosario for all the mines of the first three provinces, and San Antonio de la Huerta for the ones in Sonora. The precise description of them just made, and an inspection of their geographic locations clearly show that there is no better way of proceeding. Otherwise, there will be the inconveniences of delaying the silver on its journey to this city, of forcing the owners to take long and dangerous roads, and of causing them to face the other obstacles they have so often complained about.

Pitic, January 2, 1770   (signed)
This is a copy of the original.   de Fersén
Mexico, November 19, 1772   (rubric)

Appendix C

REPORT OF DON MIGUEL COSTANSO CONCERNING THE DISTANCE BETWEEN THE VILLA OF SANTA FE, NEW MEXICO AND SONORA, AND BETWEEN THAT VILLA AND MONTEREY [1776]

Your Excellency: I have carefully examined the two letters, that by order of Your Excellency, were furnished to me by the Secretary of the Viceroyalty, Don Melchor de Peramas. The first is from the governor of New Mexico, Don Pedro Fermin de Mendinueta, and the second is from the minister at the Zuni mission in that realm, the Reverend Father Fray Silvestre Velez de Escalante. I have prepared my report and will state my estimates in accordance with the subject matter treated in the letters, concerning the distances lying between the Villa of Santa Fe, New Mexico and Sonora, and also between that villa and Monterey. I will give calculations that reflect considerable probability and assumption in those instances which are uncertain or cannot be positively calculated. Further, I will state unequivocally the distances that can be deduced from trustworthy and credible astronomical observations, since these are the only kind of measurements that yield with an almost geometric precision the distances and relative positions of places on the earth's surface.

I consider as certain and exact the astronomical observations made at different times, and on repeated occasions in Mexico and in California by Don Joaquin Velazquez, *licenciado* of this Royal Audiencia, as well as those taken by Don Francisco Medina and Don Vicente Doz, captains of the Royal Navy sent as astronomers by our court in 1769. They were in California with the Abbot La Chappe of the French Royal Academy of Science for purposes already well known to Your Excellency.

According to those observations, Mexico is situated in 278 degrees 10 minutes [east] from the meridian of Isla del Fierro,[1] and 19 degrees 26 minutes north latitude.

Cape San Lucas, the most southerly part of California, is situated 10 degrees 36 minutes west of this capital, its longitude therefore being 267 degrees 34 minutes, figured from the aforementioned island.

The difference in longitude between Cape San Lucas and Monterey has not been astronomically observed, but our pilots compute it at 13 degrees and 11 minutes. Although the methods used by them for longitudinal calculation at sea are not very precise, they should nevertheless be given credence. In the first place, all of the pilots' diaries agree, with small variances of only a few minutes over the 13 degrees measured between Cape San Lucas and Monterey. Second, when there is only a short measure of longitude, as there is between those two points, there is little room for error in the dead reckoning of the pilots. From this, it may be said with certainty that Monterey's longitude is 254 degrees 23 minutes, and if this is not far from the truth, then the longitudinal difference between Mexico and Monterey is 23 degrees 47 minutes.

With these data, or known principles, and with the knowledge that Monterey lies at 36 degrees 44 minutes north latitude, the direct distance can be calculated between this capital and that port, and also the direction that it bears from us.

Thus, the calculated distance is 530 nautical leagues, equivalent to 705 common leagues, figuring 5000 *varas* to the league. Monterey bears from this capital approximately NW ¼ W.[2] This distance has been determined with sufficient precision and I wish that the same exactitude could be assigned to the measurement between Santa Fe, New Mexico and Monterey, or also between the same Villa of Santa Fe and one of the presidios in Sonora, like Tubac or Terrenate. Here I am forced to rely on conjecture because calculation cannot be based on completely reliable data or known principles. But acting on the assumption that Your Excellency wants

---

[1] Costansó's "Isla del Fierro" is known today as Isla de Hierro, the westernmost and southernmost island of the Canary group, located 1 degree 25 minutes west of the Meridian of Taide, on Tenerife, the more frequently used base of measurement.

[2] 303¼ degrees according to modern azimuths.

me to give my opinion in this matter, I will run the risk and speak my mind, proceeding with the consideration and prudence which would merit Your Excellency's approval. Because of Your Excellency's study and incessant labor in all branches of government, together with the reports Your Excellency has acquired concerning America, Your Excellency is a qualified judge of the subject at hand.

The road that is taken to go from this capital to New Mexico is the one that goes to Nueva Vizcaya, passing through Durango, Saltillo, Sombrerete, and Chihuahua, or the Villa of San Felipe el Real, as it is also known. This is not the most direct route, but it is used because of the greater comfort offered in following a road through populated country. The travelers and the itineraries that I have consulted agree that the direction followed is northward, with a slight inclination to the Northwest. This fact has been confirmed in many charts and maps, of which I have seen both old and new, and all of which agree in placing Chihuahua in 29 degrees north latitude. Additionally, they all confirm the city's distance from this capital at less than 400 common leagues. This is also confirmed by the observations of Engineer Nicolas de La Fora, and the map that he published of the frontier and the cordon of presidios in this Viceroyalty.

It is quite difficult not to join the unanimous vote of all travelers, especially those with some geographical training. Because of this, I do not think it can be considered presumptuous to mention these observations as deserving of some credence. I rely on them as a basis for a calculation approximating the truth.

Therefore I surmise that the traveler's distance from Mexico to Chihuahua that I estimated at about 400 leagues be reduced to the direct distance of 320 leagues, cancelling one-fifth of the road distance (undoubtedly, an underestimate of the detours). In this case the greatest difference in longitude between this capital and Chihuahua is 7 degrees 11 minutes west, so that its longitude measured from the meridian of Isla del Fierro is 271 degrees and the direction from this capital in NW ¼ N.[3]

Leaving Chihuahua for New Mexico one heads for the Rio del

---

[3] 326¼ degrees.

Norte in order to cross it as the presidio called El Paso. It is undeniable that this river flows to the east of Chihuahua and also that the route followed from that villa to El Paso must have an inclination to the Northeast. The same thing occurs on the road that goes from that presidio of El Paso to the one at Santa Fe, New Mexico. From this is can be easily inferred that the latter mentioned place is east, or rather, to the east of Chihuahua. Engineer Don Nicolas de La Fora estimates the difference at one and one-half degrees of longitude. Therefore the longitude of that villa [Chihuahua], according to La Fora, is 272 degrees 30 minutes. Its latitude, according to observations made by the same engineer, is about 36 degrees 30 minutes, which differs only slightly from older data.

The longitude of Monterey, as stated at the outset, is 154 degrees 23 minutes. That of Santa Fe, New Mexico was calculated at 272 degrees 30 minutes. Subtracting the former from the latter, a difference of 18 degrees 7 minutes results between the two spots. Making certain that I have not overlapped these calculations (I contemplate the possibility of an even greater difference in longitude between the two places), it is easy to determine the direct distance through trigonometric calculation. The result is 375 common leagues, but if the inevitable twistings of an unknown road are taken into consideration, we can well imagine the road distance to be at least 500 leagues.

Yet lacking is the determination of the location of one of the presidios in Sonora, so that with respect to Santa Fe, New Mexico, the distance between the two spots can be calculated. Considering the possibility of undertaking travel from Sonora to New Mexico, starting from the presidio of Tubac, I will seek the location of Tubac relative to some other known point, relying on the distances and observations taken in the trip made by Captain Juan Bautista de Ansa to Upper California, and also those observations that I myself made in that peninsula.

The presidio of Tubac is located at 31 degrees 40 minutes north latitude according to the map presented to Your Excellency by Ansa, which differs but little from the one attributed to it by Don Nicolas de La Fora. The distance from Tubac to the junction of the Gila and Colorado Rivers is 146 common leagues, and that spot has been observed at 32 degrees 50 minutes north latitude. From this place they [the Anza party] continued on to the mission of San Gabriel located at 33 degrees 50 minutes north latitude, and 133 leagues distant from the junction of the rivers. The longitude of Mission San Gabriel is about 257 degrees 43 minutes. From this knowledge, it can be deduced that the junction of the rivers is at 261 degrees 31 minutes and the presidio of Tubac is at 263 degrees 12 minutes.

With the longitude and latitude of Tubac known, and also those of Santa Fe, New Mexico, already indicated, knowledge of the direction between them can be established. A line between the two passes from NE ¼ E [4] to SW ¼ W,[5] more or less, and at the same time the distance is known to be 243 common leagues. In that length, the Gila River must be forded, crossing the lands of the Cosmina and Moguino Indians. And also, since it is little known country, a roundabout course is inevitable, and the distance must be taken at a minimum of 300 leagues.

The distances that I indicate are the shortest, in my opinion, that can be computed, and it would be well if this were borne in mind by anyone commissioned with the undertaking of such travels. To this end, precautions should be taken in accordance with prudence, and necessary forethought should be given to a trip that may be longer than expected. Consequently, the spirits of the people who undertake such an expedition will not be unduly weakened, and the goal proposed by the leaders will be attained.

Mexico, March 18, 1776. Miguel Costanso

---

[4] 56¼ degrees.
[5] 236¼ degrees.

Appendix D

REPORT OF PABLO SANCHEZ, SALVADOR FIDALGO, AND MIGUEL COSTANSO CONCERNING THE PROJECT OF SENDING AID TO UPPER CALIFORNIA [1795]

Your Excellency: The assistance that Your Excellency proposes to send to Upper California, in the form of colonists, artillery, and military supplies, intending to protect that land from danger of invasion or aggression by enemies, is a most difficult and arduous undertaking. Attention must be paid to the scarcity of means that can be obtained along the very extensive coastline in that province, its great distance from this capital, and the cost of such an expedition in a time as critical and calamitous as the present.

The scarcity of means consists principally in the almost complete lack of naval vessels in the South Sea. Of the two royal frigates that were in the Naval Department of San Blas, the *Concepción* sailed for the Philippines at Your Excellency's orders, convoying the ship that came from those islands to New Spain this past year of 1794. The one called *Princesa* has been reported by examiners to be in such a state of deterioration that it cannot be counted on for use on the expedition that Your Excellency contemplates.

The vessels presently stationed in the port at San Blas number only three, and they are small for troop transport.

They are a packet-boat, a schooner, and a brigantine. The tonnage of the first does not exceed 120, and that of the other two is even less. Thus, the number of troops to make up the expedition can be gauged only by the capacity of the vessels, and not by any other means. Knowing this, Your Excellency could order the commandant of San Blas to detail for you the number of persons that can go in each ship. He might also determine for Your Excellency if all three ships are available for service, or if this causes some inconvenience. Additionally, he could determine if it might be suitable to reduce the crews in order to admit more passengers. This would afford everyone more comfort and freedom, and prevent sicknesses easily contracted on long voyages when ships are crowded with people and sometimes cause the failure of expeditions. It is therefore preferable to make more trips, sending the troops in two or more voyages, rather than exposing them in the process of transportation.

According to local conditions, and in consideration of both need and means, some aid should be provided. The Upper California coast, as Your Excellency observed in the meeting to which we were summoned, is too long for us to attempt the defense of all its ports and anchorages. Because of this, only the defense of principal points should be considered, and those are the ports of San Francisco and San Diego, and the Bay of Monterey. The first, which is the farthest and most northerly of our establishments, deserves particular attention because of its natural advantages and size, and the fertility of adjacent lands. Construction was begun a year ago on a battery of twelve cannon, located on a hill that commands and defends the port entrance. Mounted on the battery are iron cannon of [blank] caliber. But although they are serviceable, there has been no one to man them since the ships of the Nootka expeditions were retired. Those ship commanders had been responsible for construction of the battery.

In order that the battery may contribute to port defense, it will be necessary to overhaul its complement of arms and equipment, and assign the required number of men to serve the guns.

Nothing has been done at the Port of San Diego, where another battery of eight twelve-pounders should be located on the point called Guijarros, opposite the entrance. But since the peninsula is barren, various difficulties may be presented in the construction of this battery; in San Diego there is a lack of construction materials, particularly wood. The battery here is going to cost more than it would if the circumstances were more favorable.

The Bay of Monterey offers little shelter to ships because of its openness and because of the distance from point to point. Only in the most westerly area of the beach adjacent to Punta de Pinos, is there an anchorage with some safety for ships. In this area alone it might be suitable to locate a small battery of four eight- or twelve-pound cannon to defend the anchorage.

These projects are designed only to provide some measure of protection from an attack by corsairs on the establishments. If the

enemy were to launch an attack with a squadron and sizeable landing forces, since Your Excellency has no matching forces to oppose
such a maneuver, they would find no resistance. In this eventuality,
the only resort of commanders of Upper California would be to
retire to the interior with the inhabitants, their possessions and
livestock, and to harass the enemy from their shelters with sporadic
cavalry detachments, if they could be put together in sufficient
number to carry out such an operation.

If the proposed batteries were to bring to bear all their artillery
at the same time, at least eight men would be needed to serve each
piece, and we calculate the total at 160 soldiers for handling the
battery. We surmise that this number can be cut in half by adapting
our defense to the attack, and to the forces of a corsair. Consequently, 32 men might be sent to the port of San Francisco, 32 to
the one at San Diego, and sixteen to Monterey.

This number should include eight or ten artillery soldiers and
corporals sufficiently trained and experienced in their work to in-
struct other troops, and they should be divided among the posts
they are to occupy. It would be very fitting if all these men were
married. If they were to be transported with their families, and if
they were to settle there, the population would be augmented, since
this is the only effective means of assuring the possession and reten-
tion of that territory.

The most careful attention should be directed towards choosing
men of good habits, hardy, and willing to work hard. They are not
only going to spend their time on artillery management (because
this would be the same as assigning them to live in idleness) ; rather,
they are to be employed under the same conditions as any other
soldier in the presidial companies of the province. This is the manner most conducive to maximum service to the king and the general
welfare of the province.

To this end each soldier ought to be allocated the daily pay and
compensation that the king normally gives soldiers of those companies which they must join in accordance with the assignment each
one receives.

It will be easy to gather such a small number of people with the
requisite qualifications by choosing them from the veteran corps of
infantry and cavalry. At the proper time, Your Excellency will

issue directions to them in order that these men may be transferred to San Blas under leadership of a few officers of the respective corps. The officers will be given the orders and instructions to be observed concerning the presence of families during the trip to that port.

This is what has seemed propitious to us in advising Your Excellency. We cannot forget the wisdom that Your Excellency has displayed in the current state of this realm, with regard to the means and resources available for executing the orders of His Majesty for the preservation of Upper California. We have avoided proposing to Your Excellency any ideas that might seem too extravagant for the state of the treasury, and we are convinced that no other more adequate solution can be devised considering the times and present circumstances. Nevertheless, we add that it appears to us essential that at all times there be two ships at San Blas dedicated solely to reconnaissance, to sail along the coast of Upper California. These ships should report every six months, if practicable, on what is happening along the coast, on the people and ships that might arrive on its shores, on establishments that any foreigners may attempt to make, etc. Only in this way will Your Excellency be fully informed and able to decree that which seems most suitable in view of events that transpire.

Mexico, July 13, 1795.   Pablo Sánchez
To His Excellency the   Salvador Fidalgo
Marquis of Branciforte   Miguel Costansó

Appendix E

REPORT ON FORTIFICATIONS

September, 1796 – Monterey
Alberto Cordoba to the Marqués de Branciforte
This is an uncertified copy [1]

In order to comply with the orders and instructions that were given to him by his director, Don Pedro Ponce, [Córdoba] proceeded to inspect the fortifications constructed to defend the port of San Francisco.

[He says] that this battery is built for the main part on sand, and the rest on loose rock; and that this part comprises the strong point of the battery, and faces to the west. There are frequent landslides in this terrain and when the rains come, it is to be feared that some of the merlons, which are made of adobe reinforced with brick, and put together with mud, may give way. Just in answering signals from ships, the firing of the cannon shakes the retaining wall. [He says] that this fort, to defend such an important point, has only two cannon facing outward, with the defect that their projectiles do not cross and as the mouth of the bay is 1,600 *varas* wide, an enemy ship could anchor without posing a threat, and out of sight of the fortifications, and could only receive two or three balls from the cannon when passing directly in front of the battery. By land, the fort is protected only by a wall of adobe with a door, and all this is located on a precipice 260 *varas* high.

[He says] that, of the thirteen pieces of artillery in this battery, there are three 24-pounders, two iron twelve-pounders, and eight bronze half culverins. Only two of the first three are serviceable because the ground is uneven. The rest of the artillery, because of its small size, cannot fire across the mouth of the bay and furthermore, even if given a target within range, there are not enough balls to be fired. The only cannon that can be used to defend the largest

---

[1] The report is a paraphrased version of Córdoba's original.

opening are poorly placed since the battery was incorrectly situated at the time of its construction, as can be seen in the attached diagram. (The said diagram does not exist in this volume.)

[He says] that the complement of the fortification consists of four troops, one corporal of volunteers, and six artillerymen. And even if the whole presidial troop should man the defenses, it would not be enough even to provide three men to a cannon. [He says] that of the 38 cavalrymen, they cannot be counted on because they are employed on escort duty.

[He says] that any enemy ship or corsair could easily disembark along the whole coast between San Francisco and Monterey, because there are plenty of good places to anchor at fourteen to sixteen fathoms at a distance of one-half mile from shore. There they could unload in small boats, as has already been done by ships of San Blas at Punta de las Almejas or Ranch of the Fathers, at Mission Santa Cruz. For this reason, he suggests augmentation of the garrison of the presidio so that men can be ready and mounted to proceed wherever they may be needed.

[He says] that repairing the fort will be a very costly operation because of the lack of labor and materials; and that a better fort should be built 260 *varas* farther back, and another on the other side of the bay at Point San Carlos. But the terrain would have to be reconnoitered first, and supplies brought in by sea, for it would be impossible by land because of the immense distance around the bay.

[He says] that for the defense of the Bay of Monterey there is a battery of ten mounted cannon of small caliber: seven eight-pounders, one six-pounder, and two three-pounders, which would only be able to defend the boats that an enemy or corsair would try to take. To defend the port they are useless because ships could anchor out of reach of these cannon. For greater clarity, he sends what he considers the necessary plans. (These plans do not exist in this volume.)

[He says] that to defend the part of the peninsula that he reconnoitered, the cost would be exorbitant, because several fortifications should be built and because it would be a good idea to have a few ships based in San Francisco ready to sail to any point in case it became necessary to evade the harassment of some corsair. From San Francisco to San Diego it is 212 leagues and there are only 300 troops including artillerymen, volunteers, and presidio cavalrymen, and as the latter are employed at fourteen missions and royal treasury ranches, there are only from two to ten infantrymen left to defend the three ports mentioned.

[He says] that according to reports, in the Port of San Diego there is a wooden shelter lined with lead with four mounted cannon to defend it. [He says] that he will pass on more information after he has reconnoitered this port personally.

Appendix F

REPORT ON THE BRANCIFORTE SITE

July 20, 1796 – Presidio of San Francisco
Córdoba to Borica

This letter is in reply to Your Excellency's letter of the sixteenth of last month asking for my opinion on the selection of the best site for the founding of a pueblo. After accompanying Your Excellency on a reconnaissance of the area immediately adjacent to Mission Santa Cruz, and in the company of Lieutenant Colonel Don Pedro de Alberni, to La Alameda and the neighborhood of the Mission and Presidio of San Francisco, I must report that the only site promising some advantages for the purpose is one on this side of the river of the said Mission Santa Cruz, or Arroyo del Pájaro.

There are good lands of all types to be found in that area, part of which can be irrigated with the construction of a dam; part with already enough water for seasonal crops; and part which can be used as new grazing lands for cattle. The area also has the essentials for construction of buildings, such as wood, stone, lime, clay for adobe, brick, tile, etc., with an abundance of water for these purposes, and with the advantage of having the sea very close by, not only for the supply of various kinds of fish, but also to facilitate, at low cost, the export of fruits and grains cultivated by the settlers. As for the increase of population in the area, it seems to me that there should be no disadvantage or prejudice to the Indians in the founding of this settlement, because the mission still has vast areas of good land for their crops and the support of their cattle.

The site called La Alameda has nothing to recommend it for the purpose. Although the terrain is level, it lacks water, not only for irrigation, but there is not even enough for the daily needs of the settlers and to meet the requirements of construction. Neither in this area is there wood, firewood, or stone, and therefore I do not consider it a good site for the purpose contemplated.

At the presidio of San Francisco and the mission of the same name, and in the surrounding area for about seven or eight leagues, there is no arable land; not even enough for a small ranch, since the earth is thin, arid, covered with sand dunes, and extremely short of water. Consequently, nothing grows there but brush and weeds. Even if good seasonal land were available to cultivate, I believe it would be difficult to work because of the constant strong wind that blows in this area. For this reason Mission San Francisco found it necessary to establish a ranch six leagues away towards the coast near the point called Almejas, in order to use some of the land there to raise the basic food for subsistence and maintenance of the Indians, since arable land is very scarce around this mission.

Therefore, if it is the intention to establish another settlement, this can be accomplished only at the site described above, next to Mission Santa Cruz, about thirty leagues from the Presidio of San Francisco, and 25 from the Presidio of Monterey.

If higher authority orders the project of founding a Spanish settlement to be undertaken, in the interests of the greatest efficiency and speed for this to be developed, it would be best for the Royal Treasury to pay for construction of the houses and to provide the settlers with all the tools necessary for all types of farming and cattle raising so that immediately upon taking possession of their land, they may quickly turn to cultivation and soon harvest grain for their subsistence. If they have to build their own houses and other necessities for the protection of their crops, it would take a year, or even two. With the shortage of people in this region, it would then be impossible for the farmer to plant his holding until the third year, which would delay him, and this delay would increase until he was unable to sell his produce in season.

Volunteer soldiers, discharged or disabled, who wish to settle in the new center, should receive somewhat more aid than the other settlers because of the merit they have accumulated during their years of service to His Majesty in an honorable career at arms.

The Indians of this peninsula have no captains or chiefs, each one living wherever is most convenient in the search for seeds to maintain himself. Therefore it will not be possible to bring in chiefs to the settlement to guarantee the loyalty of their subjects. Consequently, the only means of civilizing them is to bring in a certain number from the missions to the settlement to work and learn from the Spaniards and in time they will be able to govern themselves.

The immediate advantage of the new establishment will be the supply of grain and cattle to the Presidios of Monterey and San Francisco, and the increase of agriculture and population. As soon as there is a successful harvest, the people will be encouraged to work with more enthusiasm to promote the welfare of their children and their descendants, which coincides with Your Excellency's order.

Appendix G

DESCRIPTION AND PRESENT CONDITION OF THE
TOWN AND MISSIONS OF ARISPE WHICH HIS
MAJESTY, IN HIS ROYAL INSTRUCTIONS, HAS
DESIGNATED CAPITAL OF THESE INTERIOR PROV-
INCES; CLIMATE, PRODUCTS, AND NATURE OF ITS
LAND; ITS CHARACTER; AND CIVIL AND MILITARY
GOVERNMENT OF ITS INHABITANTS; WITH A
SHORT REPORT ON THE PROJECTS THAT HAVE BEEN
PLANNED ACCORDING TO THE POTENTIAL OF THE
LAND. [MASCARO, 1781.]

1. At the *cabezera* (or head) mission of Arispe, the town presently occupies an area of 750 *varas castellanas* in length by 400 in
width along the bank, or east side of the Gondrá River. It is on the
slope of a rocky sandhill and although it rises 150 *gemes* [1] above
river level, without much effort the town could extend to the crest.
The houses were built on two levels, or mesas, that run northeast
to southwest, divided only by a small drop that will become imper-
ceptible when the new buildings and streets are constructed. The
town is one musket-shot from the river, the waters of which enter
a poorly directed and maintained ditch at the lowest point in the
area. It could be made to carry enough water for the uses of the
townspeople with the addition of a mill, and it could irrigate several
plots that are cultivated on the lower slopes of the hill and between
the town and river if it were more evenly distributed. The town's
geographical location is 30 degrees 30 minutes north latitude, and
figured according to the routes from Mexico, it is at 266 degrees
22 minutes west longitude calculated from the base meridian on the
island of Tenerife.

2. The town is surrounded by fairly tall mountain ridges, which
extend in all directions for a distance of many leagues, and permit

---

[1] A *geme* is equal to 5.4772 inches, or 139 millimeters.

neither entrance nor exit other than the ravines formed by the rivers. To the northeast, is the Bacuachi or Chinapa valley route; to the north-northeast, the valley of the Bacanuchi; and to the southwest, the valley of the Sinoquipe. On the first route, and only as far as Chinapa, the river is forded 32 times, and a few more in the last part, all of which makes the going quite tiresome, and even more so in winter and during the rainy months.

3.   The larger and more splendid part of the scanty population of Arispe resides in the upper plain, because it is larger and because the main plaza, the mission residence, and the church are located there. The mission residence occupies the southern side of the square. It is a very tall building, but deficient in design and proportion, with a length of seventy *varas* and a width of thirty, both measurements including the thickness of the walls, which are made of adobe bricks. The roof is supported by thick beams, which is a waste of good wood. It has two sacristies, an old one and a new. The former, used only to store unserviceable furniture, is in the body of the church on the gospel side. It consists of two rooms, the first ten *varas* in length and eight in width; and the second a square six *varas* on each side, fairly even in measurement and proportion. The new sacristy was built on the same side, with a door to the chancel. It is a rectangle twenty *varas* long and eight wide. It is very dark, lighted only by a single window, which in addition to being small, is shadowed by the church walls. Its ceiling was beamed and some beams remain intact, but they have been so neglected that in one of the late rains, four beams fell and did a great deal of damage. They were repaired, but doubtless, the same thing will happen to them again and to the rest within a short time, unless they are protected with greater care from the moisture that they absorb from innumerable leaks. Materials are already being gathered for repair of the leaks.

4.   The interior ornamentation of the church is not only adequate, but rich. The high altar is consecrated to the Assumption of the Blessed Virgin, patroness of this mission, and there are two side altars, one consecrated to Our Lady of Loreto, and the other to Saint Ignatius Loyola. Both are very fine, and gilded, although they have lost some of their splendor to the dust, which is brushed off only occasionally. The ornaments, holy vessels, and other finery

are exquisite. Standing out among them is a large throne of hammered silver, a gold chalice, and an exquisite painting of Our Lady of Loreto enhanced by a silver frame.

5.   At the north end of the church, in back of the high altar, was the father minister's room. It is an old hut, 22 *varas* long and five *varas* wide. Within it is a small partition which makes an alcove, and to one side there is a kitchen, henhouse, and two corrals, but all badly roofed and almost falling apart.

6.   The Commandant General occupies the mission residence, which, though lacking in all comfort, is the best house in the whole town, and the only one with raised living quarters. It is nothing more than a large drawing-room with bedroom for consultation and another room for two servants, and at the opposite end that abuts on the church, the secretary was lodged in three tiny rooms, two of them almost useless because of their darkness, and the main one with a door to the choir, where one has to tolerate the annoyance of organ and singers practicing. The lower floor consists of six rooms, where the rest of the family are lodged in the pharmacy, kitchen, storeroom, coach house, henhouse, and two corrals that were animal shelters and could be roofed easily, and a large patio.

7.   Community housing, that today serves as barracks, is next to the church on the east side of the plaza. It is a building thirty *varas* wide and eight *varas* deep; but it is so badly arranged and in such poor repair that there is hardly room for the guards and for a couple of prisoners, who have to be kept there for lack of a public jail.

8.   The other two sides of the plaza are occupied by a number of adobe houses, low, floorless, and miserable; and only on the north face of the plaza is there a larger and more comfortable house. In it resides the only merchant in Arispe when the Commandant General arrived there.

9.   On the lower plain, the mission has a large vegetable garden, and in the center of this an edifice 28 *varas* long and six *varas* wide. There are three rooms inside with a wheat mill, built without plan, and very badly cared for, so that it is now practically useless. The mission residence, the church, and many other buildings share the defect of being built of adobe bricks, and very high, so that they are overburdened by the enormous weight of so much wood and earth

that they were being senselessly ruined until the Commandant General took the steps of ordering the roofs tiled correctly and having the interior walls plastered to protect the buildings from the rains.

10.   The rest of the town, on both levels, is a grouping of 130 little houses, positioned with no regard for order or direction of streets. Most are made of adobe, a few are stone and mud, but all are badly built with drooping roofs, dark, and dusty. Undoubtedly they would be very unhealthy, but for the climate itself.

11.   The flat, or the plain next to the town through which the river courses, is divided into two sections. The northern part measures 775 *varas* at the narrowest point. It ends at a hill that separates it from the southern section, which covers an area 1300 *varas* long and 800 wide. Gardens and other tillages have been started on the higher edges of the plain along the hills. All of this land could be brought under cultivation and guarded against flooding, which is a danger to this land when there is heavy rainfall as in 1730 and 1770; also last year, when the rains were so abundant there was hardly any undamaged area. If water from the rivers that join on the northern lowland were wisely distributed, all of the land they drain could be irrigated and made useful. The rivers would have to be dammed, or at the least, the Tahuichopa would have to be rechanneled. This would not only accomplish the purpose, but would also force higher water pressure, all the way to the level of the town's main plaza, which consequently would provide water power for driving the machinery in the mint.

12.   These two rivers are the Bacanuchi and the Bacuachi. The former rises to the north-northeast of Arispe at the foot of the Sierras adjacent to the hacienda by the same name. It flows south to a plain at the junction, and although the two rivers run parallel, they are separated by a distance. The second river originates in the northeast at a place called La Cananea. It runs through the pueblo of Bacuachi, from which it gets its name and where it receives a small amount of overflow marsh water. At a short distance before Chinapa, a spring flows into it which is called Conateboni, that erupts almost in the very river bed, and may even be a part of it, filtered through rocks in the bed. It then runs through Guepaverachi, irrigating various flat areas of land, and forming a narrow ravine between sharp precipices that follow the only trail through

these parts. It runs southwest, and at the northern end of the plain of Arispe, it joins the waters of the Bacanuchi. United, they continue southwesterly as one, watering what is really the Plain of Sonora, providing moisture for the settlements of Sinoquipe, Motepore, Banamichi, Huepac, Sonora, Aconchi, Babicora, Concepción, and Ures. Here it joins another stream, and its course twists to the west, through San José de Gracia, running through the Presidio of Pitic, after joining with the river of San Miguel de Horcasitas. These two rivers continue their joint course to Terrenate, and here the waters are dispersed through great sand banks about 25 leagues from the coast of the Gulf of California, or the Sea of Cortés, where it debouches, but only after torrents of water have been added to it from the heavy rains in this area.

13.   Despite continued hostilities of the Apaches infesting these lands, the area around Arispe is not poorly populated. Five leagues to the northeast is the pueblo of Chinapa and one league before is the Ranchería de Guepavenachi. To the north-northeast is the hacienda and pueblo of Bacanuchi. Twelve leagues to the northwest is Cocospera; at seven leagues southwest, Sinoquipe; and seven farther on, Banamichi, with several others in the general vicinity. The Presidio of Fronteras is 29 leagues to the northeast. Santa Cruz, transferred to the site called Las Nutrias, is thirty leagues north; Tucson is 65 north-northwest. Altar is ninety to the west; San Miguel de Horcasitas, newly established at Pitic, is 65 to the southwest; and San Carlos de Buenavista is 100 leagues to the south, so that the entire region is covered.

14.   The climate of Arispe is moderate in all seasons. Spring is mild and agreeable, summer is very warm, autumn rather cool, and winter is cold, with frequent snows. The winds are fierce almost all year round and usually blow from the southwest or west except in the coldest months when they are usually from the north. The climate is healthful and there are no serious diseases except in the summertime. At the beginning of this summer, there was such an epidemic of smallpox that considerable damage was done not only in Arispe and its surroundings, but also in the entire province of Sonora in general. Chronic diseases are almost unknown except for syphilis, which infects a large portion of the inhabitants, especially Indian men and women, and has even spread to the Apaches. This

terrible plague, aided by general slovenliness, failure of the Indian women to use sanitary devices, poor nutrition, poor diet, lack of means to preserve their health, and the scarcity of medicines among these unfortunate people, sends many to the grave. Nevertheless, besides the climate, the river's excellent water contributes greatly to sanitation. With its use and without any other treatment except breathing the air, many chronic sufferers have been saved, even in the hottest part of the summer, by drinking the fresh water regularly, retrieving it from various little wells that open up in the sand and imbibing it on the spot. I believe the good quality of this water stems from the many times that the course of the river changes and is filtered through underground sand, where it runs hidden for long distances.

15.   In addition to the plots immediately adjacent, this pueblo also has fourteen *fanegas* [2] of land that can be called truly fertile. Although great harvests, worthy of admiration, are produced in other parts of the northeast [*sic*] territory, especially in these Interior Provinces, the truth is that indolence and ignorance together with the turmoil in which these farmers live, are the cause of much neglect of crops and failure of the land. Wheat yields from ten to twenty to one; corn from seventy to eighty; beans from forty to fifty. Barley, lima beans, lentils, chick peas, green peas, chile, and cotton are sown in equal amounts. Fine pomegranates are produced; and also quinces, figs, apricots, oranges, lemons, limes, two varieties of nuts, prickly pears, agave, acorns, and pears, but all in small quantities for lack of dedication to agriculture. There are apples, but not good ones, although in Bacanuchi and Cuguinachi the apples are as delicious as in any other part of America. Watermelons and musk melons are delicious, and there are two harvests yearly of the latter, one in July and one in October, the second better than the first.

16.   The vegetables we have seen are not outstanding, perhaps because up to the present, there has not been an intelligent cultivator to take care of them. Sugar cane, sweet potatoes, and peanuts are very good, but above all, as far as food is concerned, the bread, meat, and water cannot be equalled in any other country.

17.   The ridges surrounding Arispe on all sides run sometimes

[2] One *fanega* equals 22.12 acres.

for leagues, and are covered with excellent grass. They form innumerable ravines and watering holes, and provide shade, so that if it
were not for the cruel and frequent harassment of the enemy, the
local inhabitants could take up seriously the raising of livestock of
which all kinds run wild as they were found in the old days.

18. Although the two rivers that water the valley are shallow
and not very long, they abound in delicious and very nourishing fish
in admirable quantities, which are caught in the short flood seasons.

19. In the nearby sierras of Mabales, Punica, Bacuachi, and
Cananea, and in various gorges near Chinapa and Sinoquipe, there
is a prodigious abundance of pine, evergreens of two species, *tascalt*
or cypress, ash, alder, poplar, Indian shot, mezquite, sour orange,
European and American figs, walnut, mulberry, sahuaro, willow,
elder, and hardwoods (like acacia, and other useful and easily
worked woods). Also, there are other bushes useless for construction but good for charcoal and firewood, such as sponge tree,
coumaru, ceiba, cat's claw, etc.

20. Herbs and medicinal plants are also abundant, and vegetables, some common to Spain and Mexico, others indigenous to
this country such as oregano, pennyroyal, sage, coriander, maiden-
hair fern, cumin, mint, peony, swallow wort, *estafiate* (which appears to be the *axenjo,* or absinthe, of Spain), maguey, tomatillo,
and grama grass; clematis, manso grass, loosestrife, tallow tree,
forage grass, Indian figs, groundsel, chayote, bur-ragweed, old
man's beard, juniper, bead tree, frangipani, sarsaparilla, senna, cal-
abash, croton, clover grass, jimson weed, and innumerable others,
all very useful.[3]

21. There is a good deal of very good turpentine extracted from
mezquite, and the country people like its flavor; they chew vanilla
beans, from which they extract a somewhat sweet and sugary juice
that does not taste very good. They use nopal cactus and the ash
plant. This and the powdered rubber called *remolino,* is burned in
place of incense and has a very agreeable odor. Above all, they
gather large quantities of *gomilla de Sonora* [4] which is much sought

---

[3] Mascaró lists several other plans, probably herbs, for which no translation or botanical reference is available.

[4] *Remolino* is a secretion of worms on certain plants, used by the Indians
as incense or disinfectant. *Gomilla,* likewise, is the gum product of insects
secreted on plants of the genus *Coursetia.*

after everywhere as its medicinal qualities become known. Mescal is also useful for its sweetness and the good liquor extracted from it.

22.   Within the pueblo there is as much rock as could be needed for new construction, and its removal is necessary in order to level the ground. In the Sierra de Mabili there is a fine granite, and much beautiful marble, and mottled jasper is from Oposura. It is so abundant that quarries can provide for the largest construction. This jasper is not only more plentiful than anything else, but also outstanding in bright colors, red, blue, straw, and violet; and in the luster it acquires when it is polished, as in a large piece I saw from that valley when it was brought in as a sample. Lime is found almost on the spot for there is plenty of it in the area, especially two leagues distant to the east where there is a little hill that could supply all the local needs for construction. Three leagues to the west-southwest, there is a good quality chalk deposit; very plentiful, white and hard.

23.   In fact, around the pueblo of Arispe, there are all the building materials that could be asked for the execution of the project, with the exception of fine woods, which although not too far distant, would be difficult and costly to haul because of the scarcity of oxen and the constant danger of attack by the enemy.

24.   Within the environs, three gold placers are known to exist: in the Bacuachi, Cananea, and Perueles ranges. In the first, nuggets up to seven marks were found, and in the last two there are many smaller ones. Currently, none of them is worked constantly because of the serious threat from the enemy and lack of water to screen the dirt, but during the wet seasons, and when it snows, some of the prospectors gather on the flanks of the Bacuachi. There are only a few of them, and they do not stop for long, due to the constant threat of enemy assault, but they take out a little gold, which pays for the costs, and last year they took out a nugget weighing seven and a half marks.

25.   Besides these placers, there is hardly an *arroyo* in the vicinity where some grains of the precious metal cannot be found. It is taken by the Indians, but currently they do little searching for it because they have other more secure means of support, and they prefer some gratuity earned in the pueblo to the uncertainty of searching for gold and the danger of persecution by the Apaches.

26.   Some other gold mines have been worked in the same area. The Santa Rosalia mine to the west-southwest of Arispe yielded gold of seventeen and a half carats, and so plentiful that some of the loads were worth a thousand pesos. This bonanza lasted the extended period of 25 years; today the mine yields no profit and is caved in, having been abandoned since the year '48. In the environs of Banamitzi there is a hill called Guija where much gold has been retrieved also, although it is of small profit today because it is not worked regularly due to the difficulties, the poor quality of the metals, and the persecution of the savages.

27.   Silver mines are found in greater numbers, and in the district of Arispe alone there are 37 old mines. The best known of them was the Espíritu Santo that yielded up to eighteen marks per load, the Rocha at twelve, and the Babicanora and others, yielding from three to six marks per load. More than twenty years ago the area was abandoned due to repeated incursions by the Apaches, who killed many of the people and destroyed the mule trains employed in carrying and working the metals.

28.   Red and yellow ocher and vitriol are also found in large quantities. There is plenty of copper, and lead and a few veins (although small), of iron have been found. On the Yaqui River there is a little hill of this metal, very plentiful and easy to work, and last year an Indian brought in a ball weighing several pounds which after a long inspection was determined to be very fine quality iron and easy to work. He said he had found it on a hill near Arispe, but it appeared to have been in the possession of somebody for a long time as it had been struck and marked, from which it is deduced that since it did not contain silver, it was discarded for richer materials. When we achieve the peace and there are plenty of settlers, it will be easier to search these hills, gorges, and valleys around Arispe.

29.   This was a larger settlement in other times. Today there are 305 Spanish settlers and mixed bloods, as well as 337 Ópata Indians, a nation which peoples a large part of this river valley and the Santa María Baserac valley. Their character in general is the same as the rest of the Indians: suspicious, lazy, indolent, superstitious, and filthy, but also long-suffering, industrious, robust, enterprising, and very brave, especially when they are with Spaniards,

whom they love. They voluntarily embraced the faith and subjugation to the crown, and although sometimes oppression made them consider slipping the yoke, or they were accused of it, the truth is that up to now they have remained loyal and obedient in the face of rigorous testing. There is nothing they value more than the honors and distinctions that reward their services to the king and their courage in the face of the enemy. If one of them marries a Spanish woman, he no longer wishes to be treated as an Indian, he scorns the work and activities of his relatives, and considers himself superior. The same thing happens to the women when they marry Spaniards. Some of them affect our dress and customs, and they act very desirous of learning the language, but when they are spoken to in Spanish, they pretend not to understand. If one made use of this inclination it would not be difficult to make them apply themselves and become industrious. The little care they have had leaves them to their old customs and abuses. Their dances are savage, accompanied by loud banging on a gourd, and their music consists of a repetition of a few notes, without expression, cadence, or harmony, and in this they differ little from the Apaches, except that the Ópatas do have some religious dances, which after a century and a half of voluntary subjugation they still preserve today. Most of their games and diversions are directed toward the exercising of muscle and agility, or in use of the bow, with which they are fairly skillful and very dedicated. Even when they arrive from their labor tired, I have seen them stop on the trail to shoot at a target. On holidays they make up hunting parties around the pueblo to kill rabbits, deer, coyotes (a kind of fox), mountain lions, and jaguars. In summary, with a school for the children and greater attention to their training, the Ópatas could be made into useful subjects and truly hispanicized, for it is undeniable that they show excellent disposition for the purpose.

30. As the settlement has gradually shrunk in numbers, there has been a corresponding decrease in the effort that is conducive to good government. At present there is only one Indian governor, one *alcalde,* and two policemen, called *topiles,* all for civil government. There is a captain, a lieutenant, and ensign, and two sergeants for the military. There is a *mador* or teacher of the doctrine, two treasurers, and two *temastianes,* or secristans for the care of the

church. Annual elections of these officials must be made among themselves without the intervention of the father minister or the police, but it is customary now for the election to take place at the entrance to the police lieutenant's house, and he makes nominations which the pueblo confirms. No sooner are officials elected than they are placed in possession of their offices without any further formalities than telling them what to do and delivering to the governor his staff of office. The officials engaged in religious activities are always selected by the father missionary and with the approval of the judge, the governor, and the alcalde, who is in charge of local economy and politics, assignment of community labor, distribution of the work, and preservation of the government and public order. The constable arrests dilinquents, and the governor or *alcalde* punishes them through the *topiles,* but when the crime is serious, the criminal is imprisoned by one or the other and they appear before the police lieutenant, who takes appropriate measures. The captain of war gives the commands in all sallies or campaigns against the enemy, and he has the power to punish cowardice, desertion, looting, and disobedience.

32.[5]   The position of *mador* is the name of the teacher of the doctrine, who instructs children of both sexes in the church every afternoon and morning and the fathers require parents to send their children at the hours assigned. On holidays, adults are instructed. The duty of the *fiscales* is to keep watch and make certain no Indian fails to attend mass on the days of obligation. They visit the sick with the *mador,* and report to the father minister on their condition, so that those in a state of grace may not die without receiving the sacraments. They accompany him when he administers them, and they bury the dead. The *temastianes* are in charge of caring for church ornaments and equipment. They clean the altars, and the interior of the temple and do whatever corresponds to the duties of the sacristan. Finally, all of them are subject to the lieutenant of the police, and he reports to the *alcalde mayor* of the province, who has no fixed residence, so that he may live anywhere he pleases within the boundaries of his jurisdiction, and he is subject to the military and political governor of the entire province.

---

[5] There is no Paragraph 31 in the manuscript.

Appendix H

Description of the Interior Provinces of New Spain [Pagazaurtundua, 1797]

Excellency: Brief Description of the Interior Provinces.

I write in reference to Your Excellency's letter of the seventeenth of last month, in which you request me to prepare a detailed description of the Interior Provinces of the Realm of New Spain by virtue of the length of time I have been stationed there. I should begin by telling you that there are five of these provinces, to wit: Sonora, Nueva Vizcaya, New Mexico, Coahuila, and Texas, all situated between 26 degrees 30 minutes and 39 degrees North latitude, and 257 degrees and 282 degrees East longitude measured from the meridian of Tenerife. Presently, all the provinces are under the orders of a Commandant General (who has an annual salary of 15,000 *duros*), and are independent of the viceroyalty of Mexico. The Commandant General has two Assistant Inspectors, each of whom has a salary of 3,000 *duros*. The Gulf of California stretches between Sonora and California, which is under the command of the Viceroy of Mexico. Considering the size of those vast dominions, the area is almost uninhabited, and its few residents, Europeans and Indians who were born and raised in those lands, profess the Apostolic Roman Catholic faith. For instruction and observance, there is at least one priest in each town.

The climate is very mild and the Interior Provinces abound in all kinds of livestock, mainly sheep. They are grown in large numbers in New Mexico and Nueva Vizcaya, as they are in Mexico and its environs. Crops of wheat, barley, and corn are raised. The first two grains mentioned are not of primary importance, but corn is most abundant, and from it, flour for dough is made for tortillas which are used as a supplement to bread (which also is made very delicious). Also from corn, *atole* is made, which is drunk in place of chocolate.

There are some shortages of fruit, but despite this, in El Paso del Norte, in Nueva Vizcaya, wine and *aguardiente* are made, and they provide one of the principal branches of commerce there. In Sonora they make a type of *aguardiente,* of several grades, that they call mescal. It is extracted from the maguey, which is called *pita* in Spain. A great deal is consumed, a fact true also of *tasajo,* which is beef jerky. But the overwhelming inclination of the people in the Interior Provinces is not agriculture, but the working of the mines. In the Sonora mines, there is a great deal of fine gold; and although Nueva Vizcaya lacks this metal, it does have very rich silver deposits, like those that have already enlarged some of the principal fortunes of Mexico.

The terror that those savage Indians called Apaches (whose exact number of tribes is yet to be authenticated) has infused in the residents there is justified. The incessant and inhuman havoc wrought in their lives and on their haciendas by the Apaches is the reason that the world does not benefit from the immense treasures buried in those mountains. These barbarians are nomad Indians inhabiting the ravines, the gulleys, and gorges of the mountains. They do not profess any religion whatsoever. They are very corpulent, especially the Comanches, who are presently at peace, and although they employ a thousand different strategies in their warfare it is their spirit that is responsible for their success. Their principal food is half-cooked horseflesh or mule, and they substitute fodder for bread. When they have no horses or mules, they sustain themselves with bull meat, deer, bison, or other wild beasts and with the brains of these animals, the women cure the hides, which they use to clothe their bodies. Those who lack even this opportunity to clothe themselves wear no more than any available kind of cloth to hide their private parts. Their principal weapons are bow, arrow, and lance, and many have muskets. On horseback, they are extremely skillful and very agile on foot too. If it were not for the incessant persecution of those valorous frontier soldiers, all of the haciendas and people in those lands would have disappeared, and the Apaches would be much closer to Mexico.

In New Mexico and Nueva Vizcaya, there is blanket and serape manufacture. They trade among themselves with these, wool and bison leggings, and furs.

In California, beaver pelts are taken in abundance. I cannot say anything about Coahuila and Texas from first hand knowledge, since I have not stepped foot in those lands. But I am of the persuasion that things are not too different there from what I have described and have always heard.

Although Nuevo Santander and the New Realm of Leon are contiguous with Coahuila and Texas, they are not considered part of the Interior Provinces. To get to these areas, one has to travel at least 300 leagues north from Mexico, all of which is free from Apaches and contains rich mineral deposits and prosperous haciendas.

I have drawn several general maps of the entire area of the Interior Provinces for the viceroys, Messrs. Florez and Revillagigedo, as well as for the Commandancy General. And since I was the only engineer there at the time, and I had other matters to attend to, I did not have an opportunity to make a copy of the map, which I would have much preferred, in order to provide for Your Excellency a clearer and more detailed description. It would have given me greater satisfaction, because I am not deceiving even myself that my report does not lack certain requisites to be expected in a paper such as this, as I am not in the best of health at the present time. In order that Your Excellency may form a more exact idea of the Interior Provinces, I am enclosing a report on the state of the military footing of the troops in Sonora, Nueva Vizcaya, and New Mexico.

May Our Lord keep Your Excellency many years. Cádiz, March 28, 1797.

To His Excellency, Mr. Luis Huet   (signed)
Juan de Pagazaurtundúa
(rubric)

Bibliography and Index

Bibliography

MANUSCRIPTS

Archivo General de Indias, Sevilla, Spain: (cited as AGI)
 Audiencia de Guadalajara, legajos 144, 242, 271, 273, 390, 416, 417, 511, 513, 516, 517, 518, 521
 Audiencia de México, legajos 346, 538, 539, 578, 579, 586, 2422, 2424, 2430, 2459, 2472, 2475, 2477

Archivo General Militar, Segovia, Spain: (cited as AGM)
 Expedientes personales – D. José María Cortés; D. Miguel Costansó, 1813; D. Nicolás de Lafora

Archivo General de la Nación, México, D.F: (cited as AGN)
 Ayuntamiento, vol. 202; Casa de Moneda, vol. 229; Historia, vols. 396, 568; Indiferente de Guerra, vols. 236, 304-A, 331; Obras Públicas, vols. 5, 6, 36; Provincias Internas, vols. 121, 169; Ramo Civil, vol. 1408; Virreyes, vol. 142

Archivo General de Simancas, Simancas, Spain: (cited as AGS)
 Guerra Moderna, legajos 2990, 3002, 3019, 3029, 3066, 3070, 3089, 3793, 3794, 3802, 3805, 3806, 5837, 7049, 7237, 7240, 7241, 7243, 7244, 7245, 7246, 7247

Archivo Histórico Nacional, Madrid, Spain: (cited as AHN)
 Diversos de Indias, legajo 464; Estado, legajos 45, 3882, 4290

Arizona State University, Tempe, Ariz: Charles Trumbull Hayden Memorial Lby., Porrúa Collection, *Elementos de Geometría* by Miguel Costansó, Mexico, 1785. 81 folios, 15 illustrations, handwritten on parchment

Bancroft Library, Berkeley, Calif: (cited as BL)
 California Archives, vols. 7, 8, 10, 14, 52, 55;
 Mexican Manuscripts, vols. 400, 401, 402;
 Manuscript Maps, Surville-Costansó, 1769, F-849-1770-C7, F-869-M66-1770-C7, 3(W)-1770-D, 12(5)-1771d-B, nos. 1-3, 18.57-1780-B, E-45-1783?-M2, Cortés, 1799, Pagazaurtundúa, 1803

Biblioteca Central Militar, Madrid, Spain: (cited as BCM)
 Aparici Collection, vols. 54, 55, 56, 57;
 Manuscripts 5-2-1-7, 5-2-4-3, 5-2-4-12, 5-3-9-5, 5-3-9-8, 5-3-9-14, 5-3-10-7, 4934

Biblioteca Nacional, Madrid, Spain: (cited as BN)
 Lafora Diary, 5963; Informe de D. Miguel Costansó, 7266; Papeles varios referentes a México (Ultramar), 19266

Biblioteca Nacional, México, D.F: Manuscript XV-2-60

Harvard College Library, Cambridge, Mass: Sparks Collection, vol. VII, ms. 98

Huntington Library, San Marino, Calif: Manuscript GA 419

Museo Naval, Madrid, Spain: (cited as MN)
 Papeles Varios, tomo IV (317); Reyno de Mexico, tomo I (333), tomo II (334); Miscelanea (485); Virreinato de Mexico, tomo I (567), tomo IV (570); Descripcion de California (621)

Servicio Geográfico del Ejército, Madrid, Spain: (cited as SGE)
 Manuscript Maps 3-3-1-10, J-2-3-96, J-3-1-14, J-3-2-48, J-9-2-c, J-9-2-e, K-b-5-8, LM-8-1-a

University of Arizona, Tucson, Ariz: University Lby., Microfilm 61

PUBLISHED DOCUMENTS

Alessio Robles, Vito, ed. *Diario y Derrotero de lo Caminado, Visto y Observado en la Visita que hizo a los Presidios de la Nueva España Septentrional el Brigadier Pedro de Rivera.* Mexico: Secretaría de la Defensa Nacional, 1946

———. *Nicolás de Lafora: Relación del viaje que hizo a los Presi-*

dios Internos situados en la frontera de la America Septentrional perteneciente al Rey de España. Mexico: Editorial Pedro Robredo, 1939

Brandes, Ray, tr. and ed. *The Costansó Narrative of the Portolá Expedition.* Newhall, Calif: The Hogarth Press, 1970

Bravo Ugarte, José, ed. *Informe sobre las Misiones (1793) e Instrucción Reservada al Marqués de Branciforte (1794).* Mexico: Editorial Jus, 1966

Cartografía de Ultramar. 3 vols. Madrid: Servicio Geográfico del Ejército, 1953

"Escalafón de 1809" [of the Corps of Engineers] in *Memorial de Ingenieros del Ejército* (May, 1908), pp. 347-51

Hoffman, Fritz, ed. "The Mesquía Diary of the Alarcón Expedition into Texas, 1718," *Southwestern Historical Quarterly,* Volume XLI (April, 1938), pp. 312-23

———, tr. and ed. *Diary of the Alarcón Expedition into Texas, 1718-1719 by Francisco Céliz.* Los Angeles: The Quivira Society, 1935

Humboldt, Alexander von. *Ensayo político sobre el reino de Nueva España.* Edited by Juan A. Ortega y Medina. Mexico: Editorial Porrúa, 1966

Kinnaird, Lawrence, tr. and ed. *The Frontiers of New Spain: Nicolás de Lafora's Description, 1766-1768.* Berkeley: The Quivira Society, 1958

Martin, Norman F., ed. *Instrucción del Virrey Marqués de Croix que deja a su sucesor Antonio María Bucareli.* Mexico: Editorial Jus, 1960

Noticias y Documentos Acerca de las Californias, 1764-1795. Vol. V. Colección Chimalistac. Madrid: Editorial Porrúa, 1959

Ordenanza de S.M. para el servicio del Cuerpo de Ingenieros en Guarnicion, y Campaña. Madrid: Secretaría del Despacho Universal de la Guerra, 1768

Ordenanza que S.M. manda observar en el servicio del Real Cuerpo de Ingenieros. 2 vols. Madrid: Imprenta Real, 1803

Portugúes, José Antonio, ed. *Colección General de las Ordenanzas Militares, sus innovaciones, y aditamentos dispuesta en diez tomos con separación de clases.* Vol. VI. Madrid, 1764

Servín, Manuel P., tr. and ed. "Costansó's 1794 Report on Strengthening New California's Presidios," *California Historical Society Quarterly,* Vol. XLIV (Sept. 1970), pp. 221-32

Teggart, Frederick J., tr. and ed. "Diary of Miguel Costansó," *Publications of the Academy of Pacific Coast History,* Vol. II, no. 4 (1911)

—— and Adolph Van Hemert-Engert, tr. and eds. "The Narrative of the Portolá Expedition of 1769-1770 by Miguel Costansó," *Publications of the Academy of Pacific Coast History,* Vol. I, no. 4 (1910)

—— and Robert Seldon Rose, tr. and eds. "The Portolá Expedition of 1769-1770: Diary of Vicente Vila," *Publications of the Academy of Pacific Coast History,* Vol. II, no. 1 (1911)

Thomas, Alfred Barnaby, tr. and ed. *Teodoro de Croix and the Northern Frontier of New Spain, 1776-1783.* Norman: University of Oklahoma Press, 1941

Torre, Ernesto de la, ed. *Instrucción Reservada qué dió el Virrey don Miguel de Azanza a su sucesor don Félix de Marquina.* Mexico: Editorial Jus, 1960

BOOKS AND ARTICLES

Bannon, John Francis. *The Spanish Borderlands Frontier, 1513-1821.* New York: Holt, Reinhart and Winston, 1970

Bobb, Bernard E. *The Viceregency of Antonio María Bucareli in New Spain, 1771-1779.* Austin: University of Texas Press, 1962

Bolton, Herbert Eugene. *Anza's California Expeditions.* Vol. I. Berkeley: University of California Press, 1930

——. *Guide to Materials for the History of the United States in the Principal Archives of Mexico.* Washington: Carnegie Institute, 1913

——. "The Mission as a Frontier Institution in the Spanish Borderlands." *Bolton and the Spanish Borderlands* edited by John

Francis Bannon. Norman: University of Oklahoma Press, 1964, pp. 187-211

——. *Texas in the Middle Eighteenth Century.* New York: Russell and Russell, 1962

Bourgoing, J.F. *The Modern State of Spain.* 2 vols. London: John Stockdale, 1808

Brinckerhoff, Sidney B. and Odie B. Faulk. *Lancers for the King.* Phoenix: Arizona Historical Foundation, 1965

Cabrera Bueno, Joseph González. *Navegación Especulativa, y Practica con la explicacion de algunos instrumentos, que estan mas en uso en los Navegantes, con las Reglas necesarias para su verdadero uso, Tabla de las delinaciones del Sol, computadas al Meridiano de San Bernardino; el modo de navegar por la Geometria; por las Tablas de Rumbos; por la Arithmetica; por la Trigonometria; por el Quadrante de Reduccion; por los Senos Logaritmos; y comunes; con las Estampas, y Figuras pertenecientes á lo dicho, y otros Tratados curiosos.* Manila: Convento de Nuestra Señora de los Angeles, 1734

Calderón Quijano, José Antonio. "Ingenieros militares en Nueva España." *Anuario de Estudios Americanos,* Vol. VI (1949), pp. 1-72

Calendario Manual y Guia de Forasteros. Madrid: Imprenta Real, 1818

Chapman, Charles E. *A History of California: The Spanish Period.* New York: Macmillan, 1930

——. *A History of Spain.* New York: Macmillan, 1918

——. *The Founding of Spanish California: the Northwestward Expansion of New Spain, 1687-1783.* New York: Macmillan, 1916

DiPeso, Charles C. *The Sobaipuri Indians of the Upper San Pedro River Valley, Southeastern Arizona.* Dragoon, Ariz: The Amerind Foundation, 1953

Domínguez Ortíz, Antonio. *La sociedad española en el siglo XVIII.* Madrid: Consejo Superior de Investigaciones Científicas, 1955

Fireman, Janet R. and Manuel P. Servín. "Miguel Costansó: California's Forgotten Founder." *California Historical Society Quarterly*, Vol. XLIX (Mar. 1970), pp. 3-19

Fisher, Lillian E. *Viceregal Administration in the Spanish-American Colonies*. Berkeley: University of California Press, 1926

Fisher, Vivian. "Key to the Research Materials of Herbert Eugene Bolton," Manuscript guide (C-B840) in the Bancroft Library, Berkeley

Holmes, Jack D. *Honor and Fidelity; the Louisiana Infantry Regiment*. Birmingham, Ala: 1965

McAlister, Lyle N. "The Reorganization of the Army in New Spain, 1763-1766." *Hispanic American Historical Review*, Vol. XXXIII (Feb. 1953), pp. 8-18

Navarro García, Luis. *Don José de Gálvez y la Comandancia General de las Provincias Internas del Norte de Nueva España*. Sevilla: Escuela de Estudios Hispanoamericanos, 1964

Priestley, Herbert I. *José de Gálvez; Visitor-General of New Spain, 1765-1771*. Berkeley: University of California Press, 1916

Richman, Irving B. *California Under Spain and Mexico, 1535-1847*. Boston: Houghton Miflin, 1911

Simpson, Lesley Byrd. *An Early Ghost Town of California: Branciforte*. San Francisco: Privately printed for his friends by Harry W. Porte, 1935

Special Commission. *Compendio Histórico publicado al cumplirse el Segundo Centenario de la Creación del Cuerpo y Dedicado a sus clases e individuos de tropa*. Madrid: Memorial de Ingenieros del Ejército, 1918, second edition

Special Commission. *Resúmen Histórico del Arma*. Madrid, 1846

Special Compilation Commission. *Estudio Histórico del Cuerpo de Ingenieros del Ejército*. Madrid: Sucesores de Rivadeneyra, 1911

Thurman, Michael E. *The Naval Department of San Blas: New Spain's Bastion for Alta California and Nootka Sound, 1767-1798*. Glendale: The Arthur H. Clark Company, 1967

Torres Lanzas, Pedro. *Relación Descriptiva de los Mapas, Planos, etc., de Mexico y Floridas existentes en el Archivo General de Indias, Sevilla.* 2 vols. Sevilla: El Mercantil, 1900

Treutlein, Theodore E. *San Francisco Bay, Discovery and Colonization, 1769-1776.* San Francisco: California Historical Society, 1968

Villasana Haggard, J. *Handbook for Translators of Spanish Historical Documents.* Oklahoma City: Semco Press, 1941

UNPUBLISHED THESES

Peloso, Vincent Charles. "The Development and Functions of the Army in New Spain, 1760-1798." Unpublished M.A. thesis, University of Arizona, 1965

Rowland, Donald. "The Elizondo Expedition Against the Indian Rebels of Sonora, 1765-1771." Unpublished Ph.D. dissertation, University of California, Berkeley, 1931

Index

Abarca, Silvestre: examinations, 43-44; engineers in Interior Provinces, 143n; promotions, 163n
Academy of San Carlos: 134
Acaponeta River: 194-95
Acapulco: naval depot, 112, 122; fortifications, 113n, 164
Aconchi: 220
Acuña, Juan de: see Casafuertes
Aduana: 197
Africa: engineers in, 38, 74; fortifications, 145
Agriculture: in Sonora, 69, 198, 221; Interior Provinces, 80, 172, 227; at Branciforte, 129-30, 213, 215; Arispe area, 154, 159, 216, 218-19; engineers' importance, 164; potential, 184; in Culiacán, 194-95; at San Francisco, 214
Agua Caliente: 197
Aguardiente: 228
Aigano: 199
Alameda, La: 129, 213
Álamos: sub-treasury, 63, 70n; mining district, 197
Alarcón, Martín de: 54
Alberni, Pedro de: California aid, 121-22; at Branciforte, 128, 133
Alcalá de Henares: 47
Alcatraz: 132
Alcudia, Duque de: Calif. defense, 115n, 117n, 121n
Alicante: 74, 87
Almodóvar, Duque de: 116
Aloy, José: 173
Alta California: see California
Altar: 198, 220
Álvarez Barreiro, Francisco: Alarcón exped., 49, 53-54; diaries,

54n; Rivera exped., 55-57; maps, 77
Andalucía: 150, 171
Antequera: 87-88
Anza, Juan Bautista de: mention, 77; overland exps., 109-11, 162, 204-05; Indian campaigns, 150
Apaches: hostilities, 28, 59, 69, 141, 147-48, 159-60, 168, 172, 181-83, 187, 194, 199, 220, 223-24, 228-29; strongholds, 149; campaigns, 149-50; Cortés' view, 180-84; dances, 225
Aragon: 75
Aranda, Conde de: 35, 113n
Architecture: church, 153-54, 160, 170-71, 217-19; civil, 142, 151, 153-54, 159-64, 168, 170, 213, 216-19
Arenas, Las: 69, 198
Arispe: mention, 55; Interior Provinces capital, 142, 167; construction at, 147; mint, 142-43; engineers in, 148, 150-52, 160-61, 163; Mascaró's report, 152-54, 159-60, 180, 216-26; co-ordinates, 216; mission, 216-17; parish church, 217-18
Arizona: 197
Army: mention, 31, 61n; reorganization, 30, 32; reforms, 31; officer corps, 32
Arriaga, Julián: 76, 82n, 109
Arrillaga, José Joaquín: 114
Arteaga, Ignacio: 112
Artillery: mention, 33, 35, 40; in California, 114, 116, 118-22, 124-25, 127, 131-32, 206-07, 210-12
Aso y Otal, Manuela de: 136

Fireman, Janet R.
  The Spanish Royal Corps of
Engineers in the western borderlands
: instruments of Bourbon reform, 1764
to 1815 / by Janet R. Fireman. --
Glendale, Calif. : A. H. Clark Co., ,
1977.
  250 p. : maps ; 25 cm. -- (Spain in
the West ; 12)
  Bibliography: p. [233]-239.
  Includes index.
  ISBN 0-87062-116-5

  1. Southwest, New--History--To
1848. 2. Spain. Ejército. Cuerpo de
Ingenieros. I. Title. II. Series.

              979/.01
              75-25210